The Noetics of Nature

groundworks |

ECOLOGICAL ISSUES IN PHILOSOPHY AND THEOLOGY

Forrest Clingerman and Brian Treanor, *Series Editors*

The Noetics of Nature

Environmental Philosophy and the Holy Beauty of the Visible

Bruce V. Foltz

Fordham University Press | ***New York 2014***

Library of Congress Cataloging-in-Publication
Data is available from the publisher.

Printed in the United States of America
16 15 14   5 4 3 2 1
First edition

Contents

	Preface	*ix*
	Acknowledgments	*xviii*
	Introduction: The Noetics of Nature	1
1	Whence the Depth of Deep Ecology? Natural Beauty and the Eclipse of the Holy	23
2	Nature's Other Side: The Demise of Nature and the Phenomenology of Givenness	42
3	Layers of Nature in Thomas Traherne and John Muir: Numinous Beauty, Onto-theology, and the Polyphony of Tradition	55
4	Sailing to Byzantium: Nature and City in the Greek East	76
5	The Resurrection of Nature: Environmental Metaphysics in Sergei Bulgakov's *Philosophy of Economy*	88
6	The Iconic Earth: Nature Godly and Beautiful	113
7	Seeing Nature: *Theōria Physikē* in the Thought of St. Maximos the Confessor	158
8	Seeing God in All Things: Nature and Divinity in Maximos, Florensky, and Ibn 'Arabi	175

9 The Glory of God Hidden in Creation: Eastern Views of Nature in Fyodor Dostoevsky and St. Isaac the Syrian 187

10 Between Heaven and Earth: Did Christianity Cause Global Warming? 203

11 Nature and Other Modern Idolatries: *Kosmos, Ktisis,* and Chaos in Environmental Philosophy 215

12 Traces of Divine Fragrance, Droplets of Divine Love: The Beauty of Visible Creation in Byzantine Thought and Spirituality 232

Notes 247

Index of Terms in Greek, German, and Latin 279

Index of Names and Places 283

Preface

And the Spirit of God moved upon the face of the waters.

—*Genesis 1: 2*

I

In a breezy, lightly forested, unassuming neighborhood in Istanbul, less than a hundred yards uphill from the powerful currents of the Golden Horn and its great ships passing by, not far from the site where the once invincible walls of ancient Constantinople were finally breached after eleven hundred years, can be found a place called the Phanar, or "Lighthouse," originally a significant region of the city, but now reduced more or less to a small compound, the Patriarchate of His All Holiness Bartholomew, 270th Archbishop of Constantinople, New Rome and Ecumenical Patriarch.[1] He is a gracious and generous man with a chest-length white beard, sparkling eyes, and a boyish smile that delights and surprises, especially given the frankness and directness of his conversational demeanor, not to mention his ecclesiastical stature. He long ago received the nickname of Green Patriarch because he was perhaps the first leader of any Christian body to speak out strongly, exercise effective leadership, and sponsor a variety of practical initiatives to address the environmental problems that may end up defining our era. And in his foreword to a recent work on the ancient Jewish and early Christian understandings of creation, he wrote something that struck me as being quite extraordinary:

> The crisis that we face is—as we all know and as we all readily admit—not primarily ecological but religious; it has less

to do with the environment and more to do with spiritual consciousness.[2]

"Not primarily ecological but religious"? *"As we all know and as we all readily admit"*? I first devoted myself to environmental philosophy in the late seventies, while still in graduate school, after a trip to the Four Corners region of the American Southwest. This splendid place was the inspiration for that epic and nostalgic genre of paintings that portrayed colorful deserts and noble Navahos to grace the Santa Fe Railroad calendars hanging prominently in every railway station, and that had made such an indelible impression on me while growing up in a small town in Kansas—one that was lucky enough to have a railroad station at all.[3] I used to lay awake at night and listen to the trains passing by on their way to this marvelous country and dream of going out there myself some day. When I finally did, "Navaho-Land," the "Land of Enchantment," was obscured and disfigured by a heavy blanket of smog issuing from the Four Corners Power Plant, whose emission standards at the time were appallingly minimal. I wouldn't let them get away with it, I thought, and chose to devote my philosophical energies to fighting the good fight—the fight against smog and air pollution. But even then, I felt that it was about something more—about land that seemed sacred, sanctified by the veneration of the Diné people, by its own incomparable beauty, and through its consecration by a half dozen generations of writers and artists. And it was for reasons like these that I chose to arm myself with tools drawn from the work of Martin Heidegger.

"Less to do with the environment and more to do with spiritual consciousness." "As we all know and as we all readily admit." It has taken three decades to reach the point at which these statements could carry much plausibility at all, perhaps even make much sense to more than a very few. But it is clearly no longer just about smog. Much more plausible, now, is Bartholomew's claim that "it is a crisis concerning the way we imagine the world, ultimately over the image we have of our planet."[4] And it is worth noting that the Greek word for "image" is *eikōn*, the word used to name those extraordinary images of Byzantine sacred art that are meant to draw together heaven and earth.

How to change the image, the icon, through which we see the world, unite our earthly life with heavenly realities? And could the earth itself even be *itself* an icon? In *Caring for Creation*, environmental philosopher Max Oelschlaeger had concluded that among human institutions, only religions have the power to motivate human action to inaugurate the needed changes on a scale that would render them environmentally effective.[5] And it was, in fact, a question posed to me by Oelschlaeger that reverberated quietly for some time in the back of my mind, and to which this book serves as the best answer I have to offer. Having read my book on Heidegger and ecology rather generously and appreciatively, he once asked me something to the effect of what we should do next, even granted that everything Heidegger had to say was correct?[6] Should we get to work teaching people how to think in a Heideggerean manner, how to master Heidegger's unusual vocabulary, how to parse the Heideggerean distinctions so well that the earth would start to recover? A question both comic and serious. Oelschlaeger himself had already provided the beginning of an answer by pointing to the major religions and their potential to change thought and attitude. Nevertheless, the talk about "stewardship" I heard coming from conventional religious quarters seemed to me tepid and quite inadequate to the task. And the idea of some new kind of new, quasi-religious consciousness being fashioned that could somehow provide more than a superficial, emotive kind of motivation for change—something like the "secular pantheism" that Michael Cohen employs to characterize the mindset of the original Sierra Club founders—seemed even less plausible.[7] I did, however, arrive at an answer to Max Oehlschlager's question, though proceeding somewhat more obliquely—addressing questions much like his, but as they had arisen from within the Heidegger book itself.[8]

There, I had argued that environmental ethics, as a branch of what was then seen as "applied ethics," was inadequate for dealing with our current environmental problems.[9] Applied ethics as a professional field rests firmly upon modernity's truncated understanding of the ethical as concerning the relatively narrow sphere of "moral obligation" and its debates over the merits of various moral algorithms, whether deontological or utilitarian, and only subsequently seeks to extend itself back out into the world of lived experience that it has, in fact, already left behind through its embrace of nature as an object of modern natural science. In contrast, Heidegger's understanding of ethics as concerning settlement or inhabitation, i.e. as concerning our characteristic manner of dwelling or "being-in-the-world," both re-captures the ancient scope of ethics and offers decisive hints for how philosophical thought could contribute toward more powerfully addressing environmental difficulties, seeing them not as resulting from incorrect moral judgments, but rather as failures in remembering what it means to dwell. And as Heidegger reminds us, often in connection with the poet Hölderlin, it is always "poetically" that we dwell, if we do dwell at all, i.e. to the extent that we still remain able to "inhabit the earth."

But if environmental ethics presupposes dwelling, and if authentic dwelling is necessarily poetic, then environmental ethics would entail an environmental aesthetic. And this is lent special credibility within the literature of environmentalism by the central role of the poetic voice—Emerson, Thoreau, Burroughs, Muir, Dillard, Berry, to name only a few—as providing the deeper tonality, the directive "drone" that gives depth and weight to the higher pitched, logistical debates about whether socialism or anarchism or capitalism, coal or gas or nuclear or solar power, or paper or plastic bags are "greener" and more "eco-friendly." But beyond this, it has been primarily contemplative, poetic voices such as these—most of whom long predate both the science of ecology, as well as voices of outright alarm in the fifties and sixties, such as Carson and Ehrlich—who have inaugurated our contemporary environmental consciousness. These, and their precursors in Romanticism and Transcendentalism—Blake, Wordsworth, Coleridge, Hölderlin, Chateaubriand, Emerson, Thoreau, and many others—not to mention visual artists such as Friedrich and Turner and O'Keeffe. And hints can be found in so many other places too, for example in composers such as Olivier Messiaen, who worked to present in his music the iconic character of bird songs, the singing of those creatures who from the earliest times have been seen to link heaven and earth, returning praise and thanksgiving for the gift of the morning or evening that its song announces. But for virtually every one of these poetic visionaries, a sense of the sacred is not only at work, but can fairly be said to predominate. Could it be that an environmental aesthetic must ultimately entail something like an environmental theology? And how could the latter be conceivable, given the withering critique of onto-theology (i.e., of theology as a metaphysical ordering of beings in relation to the highest mode of being: in relation to God as the highest entity) undertaken by Nietzsche, Heidegger, and their postmodern successors?

II

On the other hand, hadn't Heidegger himself provided a somewhat elusive hint, with his interpretation of "the holy" as Hölderlin's "name" for nature itself?[10] And only slightly less enigmatically, he had also maintained that there must be a "divine" dimension (*die Göttlichen*) to any world-structure that would allow us to save and preserve the earth.[11] But as many have wondered, along with Max Oelschlaeger, where precisely should this lead us? For John Muir, the great prose poet of the American wilderness, a realization of this sort led to his "conversion" to an inimitable life as a prophet, inhabiting what he called "so holy a wilderness" as the Yosemite and the High Sierras, a modern-day John the Baptist whose wild home was not the Judean Desert but the High Country of the American West.[12] But whence the rest of us?

It was, then, shortly after the publication of the Heidegger book that I began to discover what was, for me, a new and unanticipated continent of thought and spirituality, that of ancient Constantinople, the Byzantine East, and Russia—remarkably unexplored not only by myself, but at the time by anyone within my personal or professional sphere of acquaintance. Yet here was a very different tradition that, just as plausibly as did the Latin West, claimed legitimate descent from the ancient Greeks. Indeed, as I came to believe, far more legitimacy than what increasingly seemed like the tenuous—if not somewhat illegitimate—pedigree of Western Europe and its thought and culture in the last millennium. In short, what Heidegger had taught me to see as a hidden errancy in ancient Greek thought itself, whose later elaborations came to bear toxic fruit in the modern world, gradually came to seem instead like a default and divergence from the wisdom of ancient Greece and an unwillingness or inability to learn from its magnificent and *perhaps more authentic alternative* elaboration in eleven hundred years of Byzantine civilization (and its living presence today in Byzantine lands) and a further five-hundred-year continuation in Russia: a sophisticated and superlatively rich civilization whose very existence one could neither infer nor even suspect from the writings of Nietzsche, Heidegger, Derrida, and their successors—despite the fact that some of the most impressive expressions of Byzantine culture can be found much more recently in nineteenth- and twentieth-century Russian thought, which according to it own self-understanding, had carried on this ancient tradition after the fall of Constantinople.

Here, then, was a surprising world of thought where the tension between the visible and the invisible was not stretched to the breaking point and beyond, but rather healed and resolved and reconciled in a way that restored to both terms their weight and dignity. Here the mysticism for which I had long felt an affinity in modes of

Eastern spirituality, as well as the lure of the mythical and symbolic prior to their reduction to psychology, emerged within the heart of Western rationality not as appendages, but as its presupposition—as its deeper, richer subsoil. Here aesthetics was not sidelined and marginalized, but embedded as a central principle of epistemology and metaphysics. And here the cosmos itself, heaven and earth together, was rendered not as a springboard for feats of metaphysical athleticism, but sanctified and consecrated as our authentic home. It is, then, the gradual exploration of these new thoughts and horizons—of how they illumined the questions that had seemed dark and impenetrable before—along with my discovery of how much all this resonated to varying degrees with such seemingly disparate figures as the Andalusian Sufi master Ibn 'Arabi, the English Metaphysical Poet Thomas Traherne, and a vagabond of the American wilderness named John Muir, that I have tried to weave together in the following chapters, chapters that fit together less like a series of syllogisms than like a mosaic, or more accurately, a mosaic built of mosaics, assembled over some ten years, polished and reworked to help an overall Gestalt emerge.

Or perhaps better, a configuration of settlements grown up gradually and haltingly, initially separate locations slowly clustering around an important, sustaining feature: a bay or delta or confluence of rivers, a vast forest or a fertile savannah. Surely, however, it will not be a simple formula or grand statement, but the settlements themselves, along with the links and juxtapositions between them, that best exhibit what this source is—not as an abstracted feature of physical geography, but as a compelling, living reality for human inhabitation, in this case, one that the author is convinced needs to be much better known and more robustly occupied. Thinking itself, as Heidegger has shown us, can also be a kind of inhabitation. But for this same reason, it can at the same time serve as an invitation to dwell.

If the perspectives here allow a single conclusion, it would surely be that drawn by Patriarch Bartholomew: that our environmental crisis is ultimately a spiritual crisis, based at least in part upon philosophical-theological missteps and spiritual atrophy. Ironically, when it is put into such an abstract form, this conclusion would appear to be similar to that of medieval historian Lynn White Jr., whose work has been so influential in this area; who was, in fact, perhaps the first author to suggest that we needed a religious solution to "our ecologic crisis"; and whose ideas so many of the chapters of this book draw upon in part, while at the same time systematically working toward a point-by-point refutation of what I believe are a host of unfortunate errors embedded within them. Beyond polemics, however, each of these interwoven chapters, from one perspective or another, concludes that in the ancient thought and spirituality of Byzantium—which is still very much a living tradition in the Orthodox East, and increasingly in Western Europe and North America—and in its more distant spiritual relatives both East and West, important insights and realizations can be found that will shed an indispensible light (one for which neither scientific theory nor philosophical reflection nor practical logistics on their own can substitute) upon what it would take to sustain a happier, more salutary relation between humanity and an earth whose holiness can still shine forth like shook foil, and that we may therefore still be able to inhabit poetically.

III

Small moments can speak of much. During the first weekend in December, the Winter venue of Art Basel takes place in Miami. By some estimations, it is the largest art event on the planet. But for someone married to an artist—yet for whom an art milieu that has often left beauty behind is best taken in small doses—there are other delights to Miami during these sunny days of early December. There is, for example, "La Camaronera" in Little Havana, a hole-in-the-wall lunch counter exalted by all right-thinking diners as serving what is simply and self-evidently the best fried shrimp in the world. They are tender, sweet, and succulent—subtly flavorful, with a complexity that it may not be hyperbole to compare to some fine single-malt Scotch, and fresh enough to have just splashed up from the sea. It is that rare kind of food that elicits breathless invocations from Genesis I: Yes, creation is divinely instituted, and yes, it is indeed good! And of course there is the sea itself, pulsing from ultramarine to azure to turquoise as the clouds pass gently, casually overhead. The water temperature is perfect—still warm, but now with just the right touch of coolness to make a swim invigorating and refreshing. Drying afterward in the sun, looking out over the Atlantic toward some unimaginable shore in Western Africa, rotating the knots of the *komboskini* slowly between the fingers, and saying the Jesus Prayer in gratitude and humility at the goodness and beauty of creation, I bask in a hazy, contented sense of waves rolling in and out, and of sandpipers darting to and fro among clumps of seaweed freshly washed up onto the sand.

Ceaselessly saying the Jesus Prayer as he walks across Siberia, the Russian Pilgrim finds that all creation is transformed. "When I began to pray with all my heart," recalls the Russian Pilgrim, "all that surrounded me appeared delightful to me: the trees, the grass, the birds, the earth, the air, and the light."[13] But how easy, almost inevitable, it is for the wave or the sandpiper, the clouds or the little clump of seaweed, that one heartbreaking shade of blue, or the last savor of fried shrimp that lingers enticingly and endearingly on the tongue and in the memory, not to mention the sleek, nearly naked young bodies lounging confidently, indulgently here and there on the sand, to entice and captivate the gaze, to short-circuit the loop from creature to Creator back to creation, and instead fix our increasingly avid perception within the grip of a sorcerer's mindset, promising the possession of some alluring, indefinite treasure.

St. Athanasios, who appears in a later chapter of this book, says that idolatry begins not when we turn toward the beauty of creation—which to the pure heart everywhere sings of the beauty and goodness of the Creator—but when we turn away from it and fall instead into our own desires as if they were a cave or pit. So that climbing out, we still see them unawares, but now projected back onto creation—which by that fact becomes thick and opaque and ultimately darkened, while the gaze loses its nimbleness and lightness as it becomes congealed around some bewitching nodal point. The beheld creature is uprooted from the wisdom and beauty of the Creator, even as the fading half-life of the divine glory that we have tasted and seen and that still lingers like an aura around the things of perception now persuades us that these are themselves the true objects of adoration after all—that this is what it means to be "true to the earth," even as the earth darkens under the idolatrous gaze and each repeat performance satisfies less, making us jaded and lusting for novelty.

Natura naturans, perhaps the last hint in the West of the Eastern experience of divine energies, collapses into *natura naturata*, or rather into those one or two tokens of it that have captured us for the moment. Nor will the solemn spell of *analogia entis* do more than invoke its ancestral memory. And although he denies the comparison, Heidegger recalls us to a very similar pull—the undertow of the ontic against the ontological, pulling us away like some ebb tide until we flounder and are lost among beings.

Husserlian phenomenology, too, is sharply distinct and even dissonant, yet at the same time strangely parallel, to what will be called in this book the *noetics of nature*. And this analogy is most notable in Husserl's insistence on the necessity, and the difficulty, of the "reduction" as a portal to the noetic. Hussel's "natural attitude"—along with the ontic preoccupations that draw us into inauthenicity in Heidegger's sense—has a clear parallel to the "world" (*kosmos*) from which *askēsis* departs, only to return again renewed, reborn, not to the world as once conventionally constituted, but to the beauty and goodness of creation (*ktisis*) whose infusion with divine glories is now becoming manifest. The charitable reader, then, might even be willing to approach each chapter in this book as an exercise in seeking and practicing that return loop, the roundtrip or return home that if successful would allow us, as Heidegger's phrase intimates, to make the "leap" that will take us back where we already are—not through an attunement of Husserlian rigor nor even one of Heideggerean heedfulness, but humbly and with a purified heart, warmed by the love that it finds shining-through in the giving of the gift that is creation.[14]

Acknowledgments

In a book commending the iconographic power of the visible to lead beyond discursivity, it might appear unseemly for its author to begin by working backward from the splendid cover photograph, which speaks eloquently for itself. But if images (and the sheer visibility of creation) can direct us beyond our thoughts, this same element can also bring us back enriched to thinking, and to gratitude as well—which Heidegger has insisted is, together with remembrance, closely linked to thinking. If the *eikōn* gathers together poetically, this is at least partly in order for reflection and recollection not to explain it, but to draw from it as from a well.

To my wife, Mary, artist and professor and lover of the image, whose presence and practice have helped lure me outside the circle of my own thoughts into the open air, and who has been my faithful travel companion and indulgent accomplice across many fine and strange and wonderful lands, while enduring my sometime absence as a writer, I am lovingly indebted. To New Valamo Monastery in Finland, where she captured this photograph, looking out late from our guest-cell during one of those magical white nights of midsummer, and to those many other monasteries and monastics in Greece, Cyprus, Romania, Russia, Georgia, Palestine, Egypt, and North America (above all to Philotheou and Vatopedi monasteries on Mount Athos and to Fr. Alexios of Karakallou) who have put up with me, doubtless as a *podvig* or act of spiritual valor, in hopes that I might nonetheless learn a bit from them, I render my most heartfelt thanks. And of course there are those warm tones that the cover's designer wisely decided to reinforce, in this evocation of heaven's golden light and the silent fertility of the earth. To my parents, then, who raised me as a farm kid, close to a land where the gladdening gold of ripened wheat elevates the rich chocolate soil of the Kansas prairie to the heavens, in a world where cultivating the earth and being nourished by its beauty were never far apart—to them I will always be grateful, not least for mirroring these same faithful and luminous qualities of the place we inhabited. And to my own daughter, Emilie, and my students as well, for accompanying me on an intellectual and spiritual journey that sometimes seemed uncertain, I am profoundly indebted.

To the many colleagues and friends (too numerous to mention, but especially those at IAEP, NPR, and SOPHIA, each an island of philosophical friendship floating intact on a sea of acronyms) who have patiently and generously listened to these ideas as they took shape and critically encouraged me along the way, I am most grateful; traces of dialogue with them are everywhere present here. I am especially indebted to the readers of this manuscript, whose helpful comments and kind encouragement have been more important than they probably know. I want to express enduring gratitude to Helen Tartar, Editorial Director of Fordham University Press, who has warmly encouraged this project at every step, as well to Eric Newman, Managing Editor at Fordham, who has calmly, kindly kept it on schedule, and to Nancy Rapoport, my copy editor, who often saved me from my own errors and oversights. Finally, I want to thank Haden Macbeth for his diligent and intelligent work in preparing the index.

Earlier versions of several chapters were published, in whole or part, in other places: Chapter 2 in *Rethinking Nature: Essays in Environmental Philosophy,* edited by Bruce V. Foltz and Robert Frodeman, Indiana University Press, 2004; Chapter 4 (in Greek translation) in *PEMPTOUSIA: Politismos Epistēmes Thrēskeia,* Winter, 2006, and in *The Natural City: Re-Envisioning the Built Environment,* ed. Stephen Scharper and Ingrid Stefanovich, University of Toronto Press, 2011; Chapter 5 in *Philosophy and Theology,* Vol. 18, No. 1 (2006); Chapter 6 in *Research in Phenomenology,* 2001; Chapter 8 in *Sophia: The Journal of Traditional Studies,* Vol. 13, No. 1, Spring/Summer 2007; and Chapter 12 in *Toward an Ecology of Transfiguration: Orthodox Christian Perspectives on Environment, Nature, and Creation,* ed. John Chryssavgis and Bruce V. Foltz, Fordham University Press, 2013.

The Noetics of Nature

Introduction

The Noetics of Nature

> The sense can easily arise that we are missing something, cut off from something, that we are living behind a screen.
>
> —*Charles Taylor, A Secular Age*[1]

> As the blue of the sky and the sea changed by the hour, the thought arose, whether the East could be for us another sun-rising of light and clarity. . . . The Asiatic element once brought to the Greeks a dark fire, a flame that their poetry and thought reorder with light and measure.
>
> —*Martin Heidegger, Sojourns*[2]

I want to introduce an account of immanence and transcendence—and of the possibility of balancing the demands due to both, of being faithful to both the visible and the invisible. At the same time, it will need to trace a largely hidden dialectic—taking place in art and philosophy and theology, as well as in the deepest currents of overt schism between Rome and Constantinople in 1054—a quiet, largely unnoticed story that has nevertheless had the most profound consequences for our world, and especially for our relation to the natural environment.

The Noetics of Nature and the Old Way of Seeing

Materialism has failed. Not that it has somehow collapsed, for it is certainly thriving, no doubt more widely than before. Rather, it has failed in the transitive sense—failed, disappointed, forsaken, indeed betrayed, what it has sought to understand. This has hardly gone unnoticed in regard to the ways materialism has failed humanity, and whole libraries could be assembled of works documenting how materialistic understandings of human existence are reductionistic and dehumanizing, how they overlook what is most characteristic and important about us. But how has modern materialism failed nature as well? For it will be countered that nature essentially *is* material, lending itself "naturally" to a materialist approach? Yet the modern concept of matter—its often tacit metaphysic of a lifeless, passive, self-contained, and self-subsistent material substrate that alone is really real—differs radically from the intermittent manifestations of materialism in pre-modern thought.

The great materialist poem of Lucretius, *De Rerum Natura*, "On the Nature of Things," itself a classic of nature writing, draws us back *into* our experience of lived nature, ever attentive to the beauty of the "prescribed" and "proper" ways of nature, the ways "specific" to the things themselves, and always reverent toward the reality of the divine. In contrast, modern materialism not only fails to "save the phenomena" of nature as we encounter them in our experience—at the same time, and opposite to Lucretius, glibly assuming that things themselves *have no nature*—but actually declares war upon them through an unrelenting reductionism. For as Husserl emphasized in his *Crisis of the European Sciences*, the modern natural sciences have failed to maintain the bridges between lived reality and scientific conception. It may even be that they have burned these bridges entirely, in an unstated (and perhaps largely unwitting) attempt to annex the phenomena of the world to a conceptual framework of its own making, thus rendering modern materialism radically different from any modality in which nature has been experienced and understood, virtually anywhere outside the modern West.

Nor, ironically, has an age and a culture so solidly based upon materialistic assumptions treated nature kindly in its practices. This "strident contradiction" is brought out evocatively, using the language of sexual assault once employed by figures such as Bacon, by the contemporary Greek philosopher, Christos Yannaras:

> [Ours is] a culture which is founded on a most consistent materialism, and raises nature to the supreme level as the causal principle of existence and as a regulative authority, while at the same time it justifies and gives systematic expression to the most undisguised interventions in the laws and logic of nature, [that is,] it violates nature and treats the functional modesty of nature licentiously for the sake of an extremely risky, reckless and shortsighted utilitarianism.[3]

That materialism has failed nature is increasingly evident to many in a variety of environmental woes that have become endemic to a world that has elevated the *methodological* materialism of modern natural science and technology into an all-encompassing *metaphysical* materialism, a hegemonic materialist *Weltanschauung* that effectively rules public discourse and decision-making. For modern materialism is the metaphysical correlate of epistemological scientism, i.e., of the view that sees scientific knowledge as not just the highest, but the *only* legitimate knowledge of nature. Methodology prescribes reality, delineates its profile and parameters beforehand. Or as Heidegger put it, science is for modernity "the theory of the real."[4]

But what is theory? Before the rise of "theory" in the modern sense—and before Aristotle's celebration of what he called the *bios theōrētikos,* rendered by Heidegger as "the way of life of the beholder"—ancient *theōria* was much more than the metaphysical optics that Heidegger understood it to be.[5] *Theōria* was *contemplation,* but not the "observation" (*Betrachtung*) that fixes in place, striving to "entrap" and "compartmentalize" its *obiectum* or "object," as Heidegger contends.[6] Rather, ancient *theōria* was before this (as Heidegger perhaps intimates, in noting the manipulative sense of the Latin *contemplatio* through which the earlier Greek sense was distorted) a mystical "seeing" of the invisible within the visible, a denotation that it continues to carry to this day in the ascetic theology of Greek Orthodox spirituality, where it is used to describe the activity taking place within cliffside caves and hermitages on Mount Athos, as much as what is undertaken in the research laboratories of Athens or Thessaloniki.

Yes, *theōria* is rooted in *theasthai,* the seeing that was once associated with the theatre. But ancient drama was far from merely "looking" at what is "present at hand" (*vorhanden*) in Heidegger's sense. To be present at the theater of Eleusis and watch the divine mysteries unfold, sitting at the very place where the invisible (*haidēs,* the unseen) once emerged into visibility, and now seeing once again (*theōria*) the invisible presented dramatically within the visible—this was not a process of objectification, but an event of participation, of taking part. *Theōria* was for the ancient Greeks a special kind of attentive and "experientially engaged" seeing, closely related to wondering (*thaumazein*), as Dustin and Ziegler explain. Nor, they continue, was the *theōros* or beholder merely a curious onlooker: "A *theōros* is someone who sees (or studies), but this seeing (or studying) does not imply detachment in the way that a theoretical stance is supposed to be detached. *Theōroi* were, most commonly, ambassadors to sacred festivals who *actively participated* in the spectacles they beheld by offering sacrifices, and by taking part in the dances and games that formed an integral part of the practice of divine worship." Indeed, the Greek word *thea* "also means goddess: the *thea* of the theatre can also be read as the *thea* of 'theology.'"[7] This ancient *theōria*, then, was by no means exhausted by what Heidegger, in his lectures on Parmenides, characterizes as "the perceptual relation of man to Being."[8] Perception, rather, here extends beyond itself, seeing more than what meets the eye. Moreover, as Christos Yannaras has argued, this ancient, more participatory sense of *theōria* continues into Plato's usage as well, where it is sustained by an "erotic *astonishment*" that remains, in the words of the *Symposium*, "turned toward the open sea of beauty and contemplating [*theoron*] it." *Theōria* in Plato, Yannaras maintains, is exercised by "the eye of the soul" and "presupposes the complete experience of and participation in the beauty of what is beheld."[9]

There is surely a certain optics in this ancient way of seeing, but it is reductive neither of the visible elements that are manifest, nor of the invisible elements that elude the perception of the senses (*aisthēsis*), nor of the darkness and mystery that still remain. Rather, it apprehends the rootedness of the visible *in* the invisible: a seeing that can rightly be called a *noetics*. Here the word is employed less in the manner of Husserl and his distinction between the "noetic" and "noematic" than in the ancient usage of Parmenides and Plotinus, a luminous sense still alive in William James's *Varieties of Religious Experience*, where the "noetic" suggests "illuminations, revelations, full of significance and importance" that offer "insight into depths of truth unplumbed by the discursive intellect."[10] The word "noetic" will be used here more in the sense of Plato's divided line analogy, in which the visible becomes iconic, i.e., serves as an image (*eikōn*) for the invisible, through which the visible is in turn illumined.[11] But even more precise is the Byzantine Greek usage of the *Philokalia*, in which "*noetic*" refers to that spiritual apprehension of the invisible that can take place when the *nous* has returned from entanglement in passion-driven "thoughts" (*logismoi*) to its proper home in the heart (*kardia*), understood as the true center of human life, the "eye of the heart"—a sense that could be closer than might be expected to the German usage of the later Heidegger in words such as *Denken* and *Besinnung*.[12]

But ancient myth, and its dramatic enactment, do not inaugurate this infusion of the visible with invisible life and significance. Rather, Greek tragedy elevated and valorized in a remarkable manner the kind of "seeing" that has been normative among traditional (i.e., put privatively, "pre-modern") peoples everywhere—the salient and sustaining element of everyday experience emerging into awareness not only in ancient Greece, but among traditional peoples everywhere, and persisting even in the West until rather recently. For example, before the rise of drama, Hesiod, in his *Works and Days*, wrote lyrically to a complacent brother that success in farming requires sensitivity to, and reverence for, the sacred dimensions that sustain land and water and weather themselves—that working with the land requires spiritual eyesight, a certain kind of attentive seeing. Pindar, too, sings of these same inner realities, but as apprehended in the glory (*doxa*) displayed upon the athletic field.

Architect Robert Hale has called something of this sort simply "the old way of seeing," thus suggesting that this noetic, contemplative seeing, far from being exotic or contrived, was very much the elaboration and intensification of an accustomed, everyday mode of seeing once shared commonly. This old way of seeing, Hale argues, is rooted in our sensitive apprehension of the half-hidden, dynamic, inner patterns and orderings in the natural world that are ultimately embedded within a mystery that we learn to trust, through our familiarity with the patterns themselves and our confidence in their perennial durability. Therefore, the old way of seeing could freely design buildings and spaces that embody and exhibit this patterned, ordered mystery as the continuation of a shared conversation with nature, deriving from a common mode of attention (or *nous*) for which "sensation [itself] is a kind of visual listening."[13]

Hale believes that this old way of seeing prevailed in predominantly agrarian America until perhaps the 1830s, citing Emerson as one of the figures who documents its decline, due to the growing prevalence of a calculative, mechanistic, mercantile, and above all a mediated way of seeing. Emerson saw the rise of a "new commercialism [that] put calculation ahead of spirit," and protested in the stirring opening lines of his 1836 essay, "Nature," that "The foregoing generations beheld God and nature face to face . . . Why should we not also enjoy an original relation to the universe?"[14] Writing a few years earlier, in *Signs of the Times*, published in 1829, Emerson's contemporary Thomas Carlyle offers the same insight, unhappily noting that that we now "see nothing by direct vision; but only by reflection and anatomical dismemberment."[15] Meanwhile, writing during the 1840s, and with very different aims, Karl Marx boasts that the archaic mythic consciousness cannot seriously compete with steam engines and locomotives. And, of course, Max Weber later conceptualized this transition as a "disenchantment" (*Entzauberung*) of the world, the inevitable consequence of its growing rationalization. Seeing is replaced by calculation. Or as Hale puts it, "'Calculation' is the new way of seeing."[16] Put into Greek, *noēsis* gets displaced by *dianoia* or discursive thought as the epistemic norm, which now becomes the privileged mode of contact with the world.

But this noetic mode of seeing is not romanticist and nostalgic. Indeed, in Schiller's sense it is "naïve" rather than "sentimental." It proves itself not only by its inherent cogency and aesthetic appeal, but by its practical implications. This element is indicated by the extent to which traditional peoples have always been able to read the landscape with impressive accuracy. Wendell Berry documents this extensively for traditional agriculture, showing how a place-sensitive, sympathetic, intuitive seeing of the land has always produced better and more sustainable results than the modern agri-science expounded by the state extensions and land-grant universities.[17] And James Barr documents how Polynesian seafarers were able to sail remarkable distances, while navigating only by recourse to such subtleties as the smell of the air, the behavior of birds and cetaceans, and the taste of the sea itself.[18] The loss of this old way of seeing, then, and along with it the materialization and flattening and disenchantment of nature—the modern erosion of the living, organic *Lebenswelt* that would sustain a noetics of nature, along with the far-reaching effects of this default for the natural environment—has roots that reach back much earlier than the rise of modernity, and it is important to understand the etiology of this decline.

The Noetics of Nature in Ancient Greece

For the earliest Greeks, the noetics of nature was mythological. Their gods were radically and powerfully immanent, saturating every aspect of life and nature with what Walter Otto calls "the marvelous delineations of the divine," granting depth and weight to every aspect of nature.[19] In all that is extraordinary and intense—in whatever is striking and startling and threatens to overwhelm—the

Homeric Greeks saw a visage of divinity. Their gods were manifest everywhere in nature—in the power of the thunderbolt and the raging of the sea, in the dizzy intoxication of love, in the exhilaration of the hunt and in the quiet goodness of fecundity. In each case, the gods were "revealed in the forms of the natural, as their very essence and being."[20] At the same time, the ubiquity of conflict between the gods in Greek mythology suggests the extent to which nature was thereby seen as an arena of incoherence, and ultimately of menace, from which protection was required in the form of rituals, sacrifices, and propitiations. And yet the realm of the gods, and hence nature itself for which the gods served as a depth dimension, is not without elements of unity and coherence. Otto argues that in fact there were two competing principles of unity, one a retrograde principle, reaching back to the archaic deities, dark and disturbing and elemental, grounded in Gaia or earth, and ultimately in the Chaos which engenders all, even earth itself.[21] The other locus of unity is Olympian, heavenly, localized above all in one increasingly transcendent god, Zeus—still the fire and light striking down from heaven in the thunderbolt, but increasingly the pristine serenity of the godhead, eventually understood as wielding power even over fate or *moira*, once thought to be beyond the power of the gods.[22]

This unifying movement is taken up in Greek philosophy, for which the highest kind of knowledge, and therefore the highest moment of earthly life, still connected the visible with the invisible, was still *noetic*: the unencumbered exercise of that highest faculty of alertness and heedfulness and comprehension for which they reserved the name, "*nous*," a noun that we can render variously as intellectual intuition, immediate apprehension, *mens* or mind, or in Jacob Klein's plainer (and hence, more incisive) rendering, prepared openness, attentive consciousness.[23] Like the early Greek sensibilities described by Otto, it is a heedful awareness that is "open" to the presence of the invisible within the visible, of a kind of divinity within nature. For the connection of *nous* with transcendence, spirituality, and above all mysticism is hard to miss for anyone not already dedicated to the modern project of reading the philosophers of Greek antiquity as Enlightenment *philosophes*.

The philosophical seeing of the invisible within the visible is peerlessly exemplified by Heraclitus, who retains a robust fidelity to apperances. Eva Brann's *Logos of Heraclitus* (p. 86) puts this nicely: "Heraclitus *sees* his divinity, the Wisdom that wants and does not want to be called Zeus, in appearances." Parmenides, however, ascends from earth into the heavens like some shaman riding on a chariot of moonlight, there to fully realize in the presence of a goddess the noetic vision that he faithfully discloses to the hearers of his poem. Had he not long since been canonized as a Great Philosopher, his work "On Truth" would surely be ensconced as a Classic of Western Mysticism. For Plato, too, ascent is central and indeed characteristic of his teaching: in the *Republic* alone, there is (1) an attempted ascent from the Piraeus to Athens, (2) a mystical ascent from Hades into the daylight, (3) an epistemological ascent up the four segments of the divided line, (4) an allegorical ascent from the depths of the Earth toward the light of the Sun, and (5) a mythical ascent from earthly embodiment into that heavenly place where bodies themselves get chosen and decided. And similar ascents can be found in other dialogues, most notably the *Phaedrus*, where the soul is even granted wings to soar aloft. At the peak of the ascent is always *noēsis*: the highest seeing of what is highest. But why, in Plato and Parmenides alike, is the language that of an ascent? Why is the best kind of knowledge a "high" knowledge? From what does it ascend, if not from the earth? And whence? The movement from the dark soil of the earth to the Olympian realm of Zeus, highest of all, now seems complete.

Nor does Aristotle really break with this emphasis on ascent, as it might seem. What is intelligible for Aristotle is not material nature at all—nothing sensible, and nothing inherently dependent upon the earthly—but rather the immaterial forms, the *eidē*, that are united with the equally immaterial *nous* in the act of knowing. Thus, in Aristotle's *noēsis*, both knower and known share together in an ascent from the visible into the invisible, away from materiality toward an abstract divinity whose relation to the immanent is removed at many levels—toward the unmoved mover, understood as nothing more than ethereal, disengaged noetic reflexivity, who acts and is known only indirectly, from a metaphysical distance. And at that purely transcendent, immaterial pinnacle, should this somehow be approached, we would then find union with the divine agency itself—a dim reflection of the ecstatic unity with divinity promised by the mystery religions—with pure noetic activity ("agent intellect" in its Latin rendering) now engaged in nothing more than apprehending its own apprehension. Whether or not the specific trope of ascendancy is overtly deployed in the Aristotelean texts, it is nevertheless clear that by the time noetic activity reaches its end here, all that is earthly and particular has been left far below.

Why, then, was there no noetics of nature for ancient Greek philosophy, especially for these thinkers who held *nous* and noetic understanding in such high regard? For Parmenides, this was because nature was the deceptive fork in the road, the pathway of becoming, the road not to be taken, the way of nothingness. For Plato, *physis* has advanced to the status of what is not quite nothingness, but a marginal realm of partial truth, the ontological equivalent of Hades, the realm of *to mē on*, neither fully being nor fully non-being. The transcendent character of true being, *ontōs on*, appropriates the visible into a noetics that has from the beginning sought to leave the earth behind. Nor is this really challenged in Aristotle, for whom visibility or form (*eidos*), and the freedom from the passivity and materiality of the earth, are the criteria for true being. And the hints from Heraclitus await their revival and transformation in the experience and thought of early Christianity.

The Noetics of Nature in Medieval Thought and Experience

Into this trajectory of restless, almost obsessive, transcendence comes the peaceful image (*eikōn*) of the Nativity, a different face of being. That God has really entered into creation—not appeared by proxy like some ephemeral projection, but *come into being within the earthly*—is visually rendered in the iconographic tradition of the Christian East through *subterranean* imagery. The Eternal Logos enters substantially into creation kenotically, innocently, as a little child, represented by Mary tending to her child while inscribed deep within the Earth, in a cave, a birthplace written into the essential materiality of the Earth: Incarnation or embodiment itself is taking place within the earth, the principle of all embodiment. Home and inhabitation and immanence on the one hand, and divinity and transcendence and longing on the other, are no longer incommensurable fields of meaning, metaphysical oil and water, but are held in a serene balance.

Or we may consider the Resurrection icon, also set deep within the earth, within the realm of the invisible (*Haidēs*) where the Risen Christ is shown heroically, mystically, raising up Adam and Eve, and thus fallen and wayward humanity itself from death into life. The themes of cave and earth here resonate powerfully with the fabled Eleusian events, commending them retrospectively as types of what was to come later, but now no longer as just *mythoi* or wonderful stories: In the icon is encountered a living fusion of the visible with the invisible—the moment in which humanity and divinity, earth and heaven, are reunited within a mysterious space and time that conjoins the temporal events of Holy Saturday with the abyss of Eternity. The Invisible comes into visibility not just as temporarily or provisionally earthly, but as taking on the very nature (*physis*) or substance (*ousia*) of the earth. And in both cases, Nativity and Resurrection, light is brought into the very cave itself—the place from which Platonism had urged us to escape, if we are to find light and life—and which by that illumination is now no longer just a cave, but a holy site of illumination where it is now possible to see.

This could not be further removed from the ethereal metaphysics of classical Greek Philosophy, something we can find demonstrated in the very different senses of the term *eikōn* in Plato and in the Byzantine Christianity that transformed and arguably completed the project of Platonism. For while in Plato, especially as portrayed in the *Republic*, the *eikōn* is the lowest stage of the divided line, leading from the visible away toward the invisible, in the Orthodox icon it is just the opposite: The image here draws the invisible down into the very thicket of the visible, and this includes human beings in their fleshly, material embodiment. Indeed, in the Divine Liturgy of the Orthodox Church, the faithful are censed immediately after the holy icons because they too are flesh-and-blood icons, manifest images, of the invisible God.

The rich tradition of Byzantine and Russian iconography is rooted in a remarkable, and largely forgotten, change in modes of experiencing nature and materiality that occurred in late antiquity, and whose spiritual sensibilities have been quietly preserved and cultivated for two millennia. First, as intellectual historian Peter Brown has shown, nature was by no means de-sanctified with the rise of Christian civilization, as some have maintained; rather, places became re-sanctified—often more intensively—through remembered miracles and martyrdoms and epiphanies, as well as very tangible holy relics of many sorts, all seen as belonging not to a mythical, primordial era, but to historical time and in many cases recent memory. These holy springs and caves and pinnacles attracted pilgrims from the farthest reaches of the ancient world.[24] Second, overflowing with a new kind of spiritual energy that they were eager to cultivate, large numbers of the faithful withdrew to the deserts—first those of Egypt, and then those of Palestine and Cappadocia—and for the first time wilderness, rather than just the domesticated nature of Plato's *Phaedrus* and Virgil's *Georgics*, came to be seen as aesthetically meaningful, while providing a more adequate home for the intensive spiritual activity that was not possible within the city walls. Third, as a corollary, these wild places came to be seen as offering to the noetic eye an ancient testimony—as constituting the first, original scriptures, preceding those of Moses—and this apprehension not of dark gods, nor of a distant demiurge, but of divine order and meaning immanent and visible within creation, giving rise (long before Galileo) to the first articulation of nature as a "book" to be read, in the thought of St. Anthony the Great during the fourth century. These are some of the reasons that the Russian philosopher Pavel Florensky argues that it is only during that period, when nature was finally seen as deeply revelatory of divine truth, that what he calls "the being-in-love with creation," and what today is regarded as the "love of nature," first emerged into Western consciousness.[25]

The fourth century, in turn, opens up onto what I maintain is a classic articulation of nature in philosophy and theology, which is nevertheless not widely known in lands inheriting the Latin traditions: first because these particular developments arose largely in the Christian East—in Byzantine and Syriac traditions; and second, because to the extent they became known in Western Europe—for example, in the translations and writings completed by Scotus Eriugena—they were eclipsed and distorted by the conceptual lenses of Latin Scholasticism. The development of the noetics of nature in the Greek East, and its decline in the Latin West can in this introduction only be sampled, and in brief vignettes at that. In the fourth century, notably among the great Cappadocian thinkers, the distinction emerges between the divine *ousia* or essence, which is forever mysterious, unapproachable, and unknowable, and the divine *energeia*—energies or activity—which surround us everywhere, and are noetically apprehensible to those whose hearts have been purified.[26] Thus, a basis is laid down for apprehending the activity of the mysterious and transcendent God, encountering not the divine essence as such, but the divine *energeia* everywhere immanent in nature to those whose hearts have been healed and purified. At the same time, the great early ascetics of the Egyptian desert (and later Sinai, Palestine, Syria, and Cappadocia) are articulating a threefold mystical path: *katharsis* or purification of the soul; *theōria*, or contemplation of the divine energies; and *theōsis*, union with God and divinization of the person, body and soul. Moreover, and of decisive importance for the present study, once the soul is purified, the initial step of *theōria*, noetic contemplation, is *theōria physikē*, contemplation of nature. Thus, the same ascetics who retreated from the great cities of the ancient world to purify soul and body found themselves illumined within and through the medium of the wild nature that had become their inhabited element, reaffirming the truth of the ancient Jewish scriptures that God's first revelation is through creation. Dionysios the Areopagite, most likely a sixth century Syrian monk, even suggests that the cosmos we find around us, since it precisely *is* divine revelation intended for us, is the only world we are meant to know, thereby substituting for a Platonic movement of ascent *away* from the world the understanding that spiritual progress will consist in moving forever more deeply *into* the rich symbolism of the created order.[27] The Byzantine understanding of nature and grace, of visible and invisible, as mutually interpenetrating realms—as a marriage of heaven and earth—reaches its fullest articulation in the thought of Maximos the Confessor, a seventh-century monk, who saw the Incarnation of God as bringing to perfection the very metaphysical structure of the created order. For the eternal Logos incarnate in Christ is already expressed, partially but genuinely, in every being that is, each of whom possesses its own unrepeatable *logos*, each uniquely mirroring its Creator: each grain of sand, each blade of grass, and every leaf of every tree expresses some eternal, irreducible, ineffable meaning and revelation, offered up to a seeing that Hopkins sought for vainly in Duns Scotus, compensating instead with his own poetic brilliance. These *logoi* can be apprehended noetically through the natural contemplation (*theōria physikē*) mentioned already, every contemplation of the *logoi* of creation leading toward the divine Logos they each express, and thereby moving closer to *theōsis*—the final unification of Creator and creation. This philosophical, theological, and spiritual vision thus supports the sacramental, iconographic richness of Byzantine art, which is above all expressed in the Divine Liturgy of the Church, where heaven and earth, time and eternity, are experienced as an unfolding unity.

But this noetics of nature never really takes hold in the Latin West, with the limited exception of Celtic Christianity, whose remoteness from the centers of Western ecclesial influence helped preserve the sense of a proximity of heaven to earth even after the Norman Conquest sought to incorporate the Celtic lands into the Latin domain. The eclipse of this noetics of nature is first clearly manifest in the eighth century *Libri Carolini*, where a certain

Theodulf, attached to Charlemagne's court, attacks the Eastern veneration of icons, maintaining that the sacred, the invisible, the heavenly can better be encountered in books than in the perceptible world. But this is not just a blow against icons: It is an assault upon sacramentality as such, upon a mode of experiencing the world, and a refusal of the possibility of mystical experience, aside from what the later Latin lexicon would term supernatural infusion. And eventually, while the Byzantine dome continued to bring heaven down to earth, the Gothic spires came to point away from the earth toward a transcendent heaven, now home to a banished God—the architecture of Medieval Scholasticism. Aquinas will not allow even that last remnant of directly knowing God that Anselm had found in his ontological argument. Scotus, in turn, ontologically "flattened" the depth dimension of nature by maintaining the "univocity of being," denying the participation of the created order in the divine being, and arguing that God and creation possess the same kind of being. And William of Ockham radicalized the Scotist uprooting of the visible from the invisible by making nature a mere given, a self-subsisting artifact, whose only link to the creator lies in the capricious will of a God who chose to create this world rather than some other. The way is thus paved for Bacon, Descartes, and all the heralds of the disenchanted nature of modernity. That the beauty, goodness, and order of nature could be windows opening upon divinity, epiphanies of the divine being itself, was gradually obscured in Western thought, and remains so to this day.

The Noetics of Nature and Aporias of Late Modernity

Siphoned of the visage of the sacred, of a holy depth that we could venerate or even respect, and to which we might respond, nature in modernity is now "ours" to do with as we please. And we have seen the results. But if the traditional Byzantine understanding of creation is sound, then nature in its very essence would be revelatory of God to those who have eyes to see and hearts to heed. How, then, could this inherently sacred character have been entirely effaced through the dominance of modern thought and institutions? Of course, the eclipse of nature's "other side" has by no means been complete, and attempts to articulate a different sense of nature—as being in some way a bearer of spirit, a visible icon of the invisible—have persisted since the Renaissance, intensifying in the nineteenth century with German Idealism, British Romanticism, and especially

American Transcendentalism, with Thoreau maintaining that the only literature approximating his own experience of nature was to be found in mythological texts such as Homer, writings that evoked a world saturated with divinity. And American nature writing abounds with this invigorated sense of divine immanence in nature, from Thoreau and Emerson to Muir, to Dillard, Lopez, Nabhan, and Berry. Nor is it a coincidence that the language of this genre abounds with words like "attentiveness" and "seeing" and "wonder"—the language of ancient *theōria*. Moreover, it seems likely that the twentieth-century environmental movement more authentically traces its genealogy much more to these poetic/spiritual/noetic visionaries than to the later findings of the natural sciences. Environmental philosophy, however, has lagged behind, often remaining content with approaches, derived from "eco-systemic" concepts and methodologies, that however helpful they may be in remedying some immediate difficulty or other, reflect the very flattening of nature that constitutes the larger problem.

But the relation of transcendence and immanence as a philosophical problem remains today unresolved, and a noetics of nature is lacking, so it is hardly surprising if environmental thought has looked philosophically to the presumptive positivism of the science, even while it has listened to the voices of poets for its muse. We have considered already how the philosophical divorce of transcendence and immanence begins with the Presocratics, and notably Parmenides, for whom the transcendence of truth was so radically severed from the immanence of perceptible reality that he is forced to give two disparate accounts, the Way of Truth (much shorter in the original) and the Way of Appearing, the major part of the original poem, providing no passage between the two spheres other than the mystical chariot in which he has been privileged to journey, i.e., no bridge over the chasm, or *chōrismos*, between the illusory realm of the visible and the purely noetic realm of the invisible. Surely Plato makes important contributions to bridging this *chōrismos*, notably his understanding of *methexis*, or participation, as does Aristotle with his understanding of being as *energeia*, or actualization—later to be employed in the Byzantine tradition as the key to bridging transcendence and immanence through an understanding of *energeia* as referring to the divine energies infusing and upholding and rendering meaningful all of creation. In some measure, *methexis* and *energeia*, in their Latinized forms—in the Neoplatonic theory of forms that develops with Augustine and the Aristotelean analogy of being in Aquinas—hold together heaven and earth during the Western Middle Ages as well, until the collapse of the medieval synthesis with Scotus and Ockham, whose legacy is a cosmos that is *no longer ontologically capable of bearing the weight of divine epiphany*—impermeable not only to the divine energies, of which the Latin West had always been suspicious, but even to the "forms" as Scholasticism had come to understand them, as well as any desperate *analogia entis* between Creator and creation. The stage is set for the modern sense of creation as brute *factum*.

Remnants of ancient solutions to the relation of transcendence and immanence intermittently resurface in Renaissance philosophy, under the rubrics of Platonism and hermeticism, even as the art of the time embraces a "realism" that presents nature not as porous to transcendence, but as opaque and self-sufficient.[28] Gradually, modern science moves from seeing the "laws" of nature as insights into a transcendent wisdom and perfection immanent to the world, to a purely operational, "empirical" study of regularities that happen to be as we find them to be, and which we are free to exploit as we please without the hindrance of transcendent considerations. The antipathy between the two "paths" becomes the leitmotiv of modern philosophy through an ingenious array of forms within the parallel universes of British Empiricism and Continental Rationalism, followed by Kant's purported resolution through an asymptotic approximation to the noumenal by means of "practical reason," now understood in a thoroughly calculative mode. In the wake of the *Pantheismusstreit*, German Idealism makes a last, great, metaphysical effort at retrieving the transcendent within the immanent, only to conclude with the hegemony of the secular in Hegel, seen as the irreversible kenōsis and final actualization of transcendence—and hence as the triumph of perfectly fulfilled immanence. Hence too, the resistance movement of Kierkegaard and Nietzsche to pursue an indirect resurgence of the transcendent (respectively, God and Eternal Return) back into immanence, precisely by means of radicalized commitment to immanence itself (respectively, existence and affirmation of "time and its 'it was'"), each of them projects designed for solitary execution.

With the rise of phenomenology, the terms "transcendence" and "immanence" themselves begin to be used explicitly by Husserl, even if in a strangely truncated manner, and developments in phenomenological philosophy eventually give rise to what is becoming recognized as a major, unresolved aporia in the postmodern thought of the late twentieth and early twenty-first century, one that is often associated with the "return to religion" (Vattimo) or the "theological turn" (Janicaud), but that extends much farther into postmodern concerns. On the one hand, we find a number of postmodern thinkers emphasizing radical, irreducible, intractable transcendence, often under the rubric of the Other. Levinas and Derrida are examples of this tendency. On the other hand, others embrace an equally radical immanence, either of the Hegelian mode, in which the transcendent has been fully actualized and in the literal sense of the word, rendered mundane—or else of the Spinozist or Nietzschean modes, in which an attempt is made to render radical immanence itself venerable in some new way. Certainly Deluze, for whom the life of "pure immanence," "absolute immanence," ultimately "attains a sort of beatitude" is the preeminent example of the latter tendency.[29]

John Panteleimon Manoussakis has made the overcoming of this polarity a focus of his work in philosophical theology, looking first to the work of Richard Kearney "to steer a middle path between Romantic hermeneutics (Schleiermacher), which retrieves and appropriates God as presence, and radical hermeneutics (Derrida, Caputo), which elevates alterity to the status of undecidable sublimity."[30] But it is in his later book that he works most directly toward a resolution of the problem, which he now characterizes not epistemologically, but as a "metaphysical dilemma":

> Either an unknowable, imperceptible, wholly other God, or a conceptual, and therefore equally fleshless Idol: either *Gott* or *Götze*. In the modern past, metaphysics was content with the latter, that is, with the *idea* of God (see, for example, Descartes' *Third Meditation*). It is that contentment that has been called ontotheology. Phenomenology, on the other hand, all too often rushes toward the former, mesmerized by the lure of the otherness of the other like a butterfly bedazzled by fire.

And although his immediate concerns are somewhat different than those I am addressing, Manoussakis's resolution is the same as I hope to outline here: a "third position" centered upon the paradox of the visible "icon of the invisible God."[31] Parallel to this, and working from within the strongly Western framework of Radical Orthodoxy, James K. A. Smith (along with colleagues such as John Milbank, Graham Ward, and Catherine Pickstock) has concerned himself with the same false dilemma between transcendence and immanence in modern and postmodern philosophy—between "Gnostic dualisms, which opposed the transcendent to the immanent, and atomistic materialisms, which flattened the world to mere immanence"—arguing instead for a solution based upon a retrieval of *methexis* and the *analogia entis*, an outcome that I will argue proceeds in the right direction, but by remaining too uncritical of Western metaphysics, is incomplete.[32] Charles Taylor refers to the present-day "emptiness of the everyday" as a "malaise of immanence" resulting from the "eclipse of transcendence," while Guido Vanheeswijck complains that the "openness to transcendence" in current French philosophy "is accompanied by an aloofness to filling up the contents of this transcendence."[33] Many other examples could be cited.

But what is important to note here is that both radical transcendence and pure immanence "flatten" the depth dimension of the natural world, rendering it two dimensional, as I will argue in Chapter 2. Milbank, Ward, and Pickstock have put it nicely in the Introduction to their manifesto volume, arguing that "if there is only finite matter, there is not even that, and that for phenomena to be there they must be more than there."[34] But if this is the case, then a noetics of nature (toward which this volume would be no more than a prolegomena) would not only offer a philosophical context for environmental thinking, but would constitute an important step toward resolving the terms of the aporia between a world utterly bereft of a God whose distance and height and icy aloofness become a kind of "impossibility" (Levinas, Derrida), and a world of "asphyxiating" immanence (Deleuze, Foucault): "the pseudo-dilemma between an either *sur*-real or *sub*-real God."[35] And given this aporia, we are situated between a *de jure* materialism of radical immanence and a *de facto* materialism of total transcendence, with both leading to the same result of a positivistically conceived "natural environment" that is impervious to any meanings beyond those accruing to a subjectively projected, recreational resource—or else those implied by its possibilities as a more practical resource, whether for raw materials or for scientific research. Moreover, even among those for whom the fact of this philosophical aporia is itself proof of the contemporary irrelevance of philosophy, the default ontology of nature nevertheless becomes the uncritical materialism to which more popular renditions of modern thought and practice have been leading all along.

But doesn't Heidegger's philosophical work serve, among other things, as a powerful and perhaps even exemplary noetics of nature, finally drawing together the realms of transcendence and immanence and resolving the aporia? For example, aren't themes such as " the thing" and "the fourfold" as they are explored in Heidegger's later thought examples of a philosophical reflection upon our encounter with the inner meaning, the poetic "logos," of simple, everyday realities that surround us: trees and mountains, the sky and the earth? And what of his essays on the poets—Hölderlin above all, but also others such as Rilke, Trakl, Hebel, and George—and upon the poetic discourse that concerns things earthly and heavenly? But although something like a noetics of nature is often approached in various hints and intimations and divinations, it is everywhere deferred and nowhere explicated: In his later work as much as in the early writings, transcendence is approached only mantically. It is only suggested in his reflections on poetry, which rather single-mindedly draw toward purely ontological issues and conclusions—in fact, away from things, in a counter-movement to the poets themselves, with regard to whom Heidegger posits an abyss between their work and that of the philosopher. It can perhaps be conceded that rather than a noetics, Heidegger outlines something like a "poetics" of nature. But given the exceptional status of the poet, whose prophetic vision has little connection with our own everyday experience of the world; given the apocalyptic orientation that he emphasizes in the work of poets he takes seriously; and given also the "abyss" Heidegger claims between the poetic and the philosophical, it is not clear how helpful such a "poetics of nature" would end up being. We do, perhaps, find here and there a noetics of things-made, of peasant shoes and Greek temples, of pitchers and coffins and prayer corners in peasant cottages; but apart from their more poetic approach, it is unclear how these considerations would differ fundamentally from the "eidetics" of artifacts to which Plato so often seems to retreat in his dialogues, when difficulties arise concerning non-artifactual *eidē*. Where is the "natural environment" in all this? In an unusual reference to something natural—in *What Is Called Thinking?*—we hear that philosophy has not yet been able to let a tree actually *be* a tree. But in fact, does Heidegger do more here than problematize this fact, himself leaving the tree behind, quite as much as those he criticizes? Or in the same Parmenides lectures noted earlier, Heidegger seems at last to draw near to just such a contemplative relation to things, only to quickly drown it in the chilly waters of ontology:

> To think Being does not require a solemn approach and the pretension of arcane erudition, nor the display of rare and exceptional states as in mystical raptures, reveries, and swoonings. All that is needed is simple wakefulness in the proximity of any random, unobtrusive being, an awakening that all of a sudden sees that the being "is."[36]

To paraphrase Hegel, the thing itself here gets swallowed up from the start into the ontological night in which all cows (or trees and mountains, or sky and earth) are black. The "isness" of the tree, to draw upon the usage of Meister Eckhart to whom the later Heidegger so often turns for guidance, overwhelms the tree itself. Nor should we overlook the dismissive identification of the mystical with "raptures, reveries, and swoonings," as if the latter typified attunement toward the noetic and the transcendent. And the "simple wakefulness" commended here, in contrast to both the Hesychastic *nēpsis* (watchfulness) and the Buddhist mindfulness (*sati*), seems already resolved upon what it is alert *for*, a temporary expedient to help us "see that the being 'is.'" If it generates a certain wonder, it leaves much more as merely taken for granted. That the chambered nautilus or Half Dome "is" is indeed (in a purely ontological sense) remarkable, but no more and no less than is the case with anything else, a pencil or a mailbox or anything else "random" or "unobtrusive." But there are other remarkable things to be seen about the nautilus than the mere fact of its being—there are inner depths or *logoi* that they harbor. And only if we see these things first, do we *then* see that the fact that the *nautilus* "is," is in fact, *more* remarkable than the fact that the desk chair is. It "says" something more. It means something much more. If it is indeed an occasion for wondering that anything at all can be, how much more wonderful that this reality *that is itself wonderful* can be!

Only in some of his autobiographical reflections, such as *The Country Path* (*der Feldweg*) and *Why We Remain in the Provinces*, mediations that in the language of his earlier thought Heidegger would have called "factical," do we get anything approaching a noetics of nature, and this only in passing. Why, then, such a lacuna? Here it must suffice to note that beginning with his decisive reading of Luther, Heidegger became convinced that what he, along with Luther, took to be "primal Christianity" had provided the model for a radically new form of life (a "factically" executed life) not just by a rejection, but by deliberately engaging in a "destruction"—Luther's *destructio*, rendered in German as *Destruktion*—of pagan "contemplation," exemplified not by the religious ritual, nor even less by the pastoral poem, but by Aristotle's remarkably arid understanding of the noetic pinnacle of the philosophical life, as a kind of intellectual intuition that had itself as its own object.

Subsequently, Heidegger rejected anything that might have seemed like a noetics as necessarily constituting a retreat from life and action and existence, a regression back into pagan self-indulgence. And from the time of his early critique of the ontology of *Vorhandenheit*, or "mere presence," to his later deconstruction of onto-theology—both of which he saw as evoking the contemplation that distracts from "life" and "action"—Heidegger characteristically retreats into a resounding silence (curiously similar to the positivistic reticence prescribed by the early Wittgenstein regarding the "mystical") whenever occasions arise where he might be expected to speak about theological matters, *or* about the "nature" that lies around him, in a philosophical matter, lest he fall into the *theologia gloriae*—or what we might call the noetics of God's glory (*doxa*) in nature—which Luther had so brashly condemned in his 1518 Heidelberg Dissertation.[37] That there could be, and could have been, another kind of *theōria*, a *theōria physikē* or contemplation of nature that not only remains within the "factical" but intensifies it, fulfills it, indeed radicalizes it, Heidegger seems never to have considered. But all this will require more detailed examination in later chapters of this book, especially at the beginning of Chapter 3.

Thus, given its absence even in the work of Heidegger, a philosophical retrieval of the noetics of nature, especially in its most ancient and integral expressions, would be a task of great urgency in an age in which our alienation from nature presents us with consequences that can no longer be ignored. Technological fixes will not suffice, for we need changes that acknowledge what has actually been lost: not technical mastery, but the noetics of nature, and even the "old way of seeing" that would serve as its perennial, preliminary ground. Thomas Traherne, the seventeenth-century English poet, wrote: "You never Enjoy the world aright, till you see how a [grain of] Sand Exhibiteth the Wisdom and Power of God." That is, until we see nature as a splendid gift from its Creator, meant to entice us into enjoying it—and thus into enjoying the divine beneficence itself—and in the refusal of which, we sin against divine love, dishonor the gift of nature, disfigure our own deepest humanity, and leave environmental devastation behind us.

How we treat the natural world around us—"the environment," as environmental science calls it, "creation" (*ktisis*) and "cosmos" (*kosmos*) as the ancient Jews and Greeks and early Christians called it—follows effortlessly, quietly, but inexorably from the way we see it. From the way we conceive it. *From our relation to it*. And from the way we speak about it. The poet and agrarian writer Wendell Berry states this with eloquence:

> The problem, as it appears to me, is that we are using the wrong language. The language we use to speak of the world and its creatures, including ourselves, has gained a certain analytical power (along with a lot of expertish pomp) but has lost much of its power to designate *what* is being analyzed or to convey and respect or care or affection or devotion toward it. As a result we have a lot of genuinely concerned people calling upon us to "save" a world which their language simultaneously reduces to an assemblage of perfectly featureless and dispirited "ecosystems," "organisms," "environments," "mechanisms," and the like. It is impossible to prefigure the salvation of the world in the same language by which the world has been dismembered and defaced.[38]

Modern science and technology stand as the culmination of two and half millennia of Western thought and experience. But if there has been a profound errancy in this history regarding the relation of the visible and the invisible (as even Nietzsche insisted, although seeing only the outcome and wreckage, he did not fully grasp the character of the problem) and thus between nature and our way of comprehending it, then the very means that we use to understand our environmental problems could themselves be deeply implicated in the factors that generated the problems themselves. The following chapters, then, will maintain that the retrieval of something on the order of a "noetics of nature" is required for us to restore a salutary relation between ourselves and the natural world that, as we are continuing to learn, is in increasing danger.

But despite their unquestionable importance, and despite their growing urgency, purely ecological issues are by no means all that are at stake. What also remains at issue is put well by S. H. Nasr: "One must ask how can one destroy through purely human agency the vessel, the form in which the Sacred has manifested itself, without further destroying access to the Sacred?"[39] The argument of this book, then, is that in truth, these are not two different issues at all, and that we thus cannot address either of them well without also addressing the other.

CHAPTER 1

Whence the Depth of Deep Ecology?

Natural Beauty and the Eclipse of the Holy

I

> From a deep dream I woke and swear
> The world is deep,
> Deeper than day had been aware
>
> *Friedrich Nietzsche, "The Other Dancing Song,"*
> Thus Spoke Zarathustra

Despite their pretensions to depth, dreams are affairs of the surface. Which surface? The dreaming consciousness enjoys a constant motion, but always a lateral rather than a vertical movement, one that never arrives at its own denouement since the advent of finality is inevitably the moment the dreamer awakes—to the deeper breathing of relief or of longing, but awakes outside the dream even as it is shed off, just as the swimmer emerges dripping-wet from the surface of the water, plunging only now upon awaking into the truer depths of what transcends the self. The motion of the dream, however relentless it may appear to the dreamer, is unreal, goes nowhere, is a motion of Tantalus, always a movement in place. And as often as not, the lateral character of the dream is perfectly evident, sometimes even evident within the dream itself, as if it were hyper-text displayed on the screen of a computer monitor, where the links patently lead no place but to more links and these to yet further links, flashing one after the other across the surface but arriving nowhere. The dark spots on a banana peel become markings on the shell of a beetle that turn into a person who once stood behind me on a bus in some other time and who now appears at my door. But what, again, constitutes the surface of the dream? Surely nothing other than the dreaming consciousness itself. The dream logic tolerates no unobstructed apertures to anything outside itself, exterior to itself, deeper than itself. The dream is, as Descartes saw with nightmarish clarity, a solipsistic event, even (and most impressively) when some other event outside the dream (a noise or a change in ambient temperature) gets incorporated seamlessly into the self-driven logic of the dream-life, one that can never admit of an outside.

In "Helen's Exile," an essay originally dedicated to René Char, poet of nature and its stern beauty, Albert Camus maintains that Western humanity has exiled nature and beauty and the sense for limits that arises from contact with them, while encapsulating itself within the dark dreams of history:

> "Only the modern city," Hegel dares to write, "offers the mind the grounds on which it can achieve awareness of itself." We live in the time of great cities. The world has been deliberately cut off from what gives it permanence: nature, the sea, hills, evening meditations. There is no consciousness any more except in the streets because there is history only in the streets, so runs the decree . . . History explains neither the natural universe which came before it, nor beauty which stands above it. Consequently it has chosen to ignore them.[1]

Cities and history. And now beyond these partitions, texts and signs—enclaves that can be even more stubborn in their resistance to anything outside them, anything exterior to them, anything "deeper than day had been aware." After Camus, postmodern culture generates its own endless hyper-text, yet with the innovation that it wishes not simply to ignore, but to refuse outright anything outside the text as a proper matter for thought or experience. For the postmodern condition, the word "deep" is at best an emblem of naïveté, and more likely an indication of disloyalty and hostility to the culture of post-modernity, a token of ill-will and treachery—or perhaps even, worse yet, a symptom of "essentialism." "Culture," now defined less historically than semiotically, becomes in turn the preferred term for reality itself. This is, indeed, still a humanism, but a lonely humanism in which even a shared humanity is disdained, in favor of an archipelago of cultural gulags, various overlapping "cultures" within which we are each, and collectively, expected to find our proper residence. And from which we can only dream of waking.

In the last quarter of the twentieth century, then, deep ecology stood as one of the few intellectual movements to withstand the sophisticated humanism of post-modernity, to resist the socio-cultural constructivism that is now creeping even into environmental philosophy, and to insist that nature was real, and that "shallow" and "deep" had legitimate referents.

II

> The Heideggerean literature at step x=+2 will be characterized by some of us as murky rather than deep, or at least both murky and deep.
>
> *Arne Naess*[2]

What makes deep ecology deep? What is the dimension of depth for deep ecology that will keep it from being shallow? In an article called "Deepness of Questions and the Deep Ecology Movement," Arne Naess reflects on this question of the "depth" of deep ecology. He begins by acknowledging one sense of depth, according to which it is the deepness of the questions it poses that constitutes the depth dimension. Deep questioning is one that progressively, step by step, questions answers—showing at every step that each answer can in turn be seen as provisional and shallow relative to another, deeper question: "Persistent questioning leads to deeper questions."[3] But there is more to this than just the propositional sense in which a premise is deeper than a conclusion, for as we proceed in uncovering and questioning presuppositions, we attain an overall net depth—i.e., our questioning becomes not just deeper in the relative sense, but becomes as such more "profound." This sense of "profound," in contrast to the "deepness" of "premise/conclusion relations," Naess lets "refer to nearness to philosophical and religious matters," yet surprisingly adds: "the latter term [i.e., 'profound,' so characterized] I leave unanalyzed."[4]

But if this profundity that is more important than propositional depth is left undefined, how can we avoid the danger of becoming shallow in our very questioning, especially given the danger, at least suggested by Naess, that continued questioning may lead not to greater "profundity" at all, but to "certain kinds of diversionary steps or side-tracking maneuvers," i.e., into questions of professional philosophy that, if taken up as an end in themselves, lead not to greater overall depth but into conceptual shallowness and sophistry. Perhaps one answer can be found in Naess's claim that "'deepness' must include not only systematic philosophical deepness, but also the 'deepness' of proposed social changes."[5] The depth of questioning that would be "profound," then, would lead to social change that—in the words of point six of his deep ecology platform, to which Naess refers us as an index of deep social change—"would affect basic economic, technological, and ideological structures. The resulting state of affairs would be deeply different from the present."[6]

But should we not question more deeply here, too, in hopes of arriving at an element that is itself more "profound"? What kind of difference would constitute a "deep" difference from the current state of affairs? And in what direction should these changes move? How do we know that the differences at which we are aiming will not turn out to be shallow, that the specific changes for which we are struggling will not be discovered to make no deep difference at all? How do we know, for example, that our attempts to bring about a deeper respect for nature by establishing carefully protected parks will not end up transforming the protected areas themselves into specimens and mere objects of curiosity, vanquishing the very wildness we aimed to protect? How do we avoid the self-deception of the dreamer, whose belief in waking is just another part of the dream, a sophisticated gambit that allows us to linger in the shallows? Can we dispense with a philosophical elucidation of this depth dimension that would be substantive, and that would provide us with some criterion for knowing when the questioning of deep ecology, or of environmental philosophy as such, is getting deeper into what Husserl called *die Sache selbst*, the very "thing itself," and this must surely mean *nature* itself? And for knowing when it is merely superficial and self-enclosed? What is the depth of nature—and so too, what is the ground in which deep ecology is properly rooted? As Heidegger observed of Descartes's project of radical questioning, there is always a special danger that, as we seek the roots or fundamentals, we will overlook the very ground that nourishes those roots—a ground that just because of its axiomatic—that is, its sustaining and ultimately earthly character—will necessarily appear murky to propositional analysis.

This chapter, then, will investigate a certain depth of nature itself, a depth that is hardly unknown, yet which remains largely unarticulated philosophically. This will be defined first by means of an aesthetics that is proper to nature, and which will lead to the conclusion that natural aesthetics is philosophically prior to the aesthetics of art. These considerations will, in turn, lead to the phenomenon of a kind of integrity or "holiness" in nature, without which this depth of nature cannot be understood. Finally, this holiness of nature will serve as a basis for unifying some of the central issues of environmental philosophy, as well as showing why additional issues should be seen as more central. It will, then, try to pursue a perhaps "deep and murky" path through some of those very "philosophical and religious matters" that can help elucidate that ground in nature itself within which environmental philosophy needs to be rooted, if it is not to end up being shallow and merely academic—i.e., if it is not to digress into a series of "diversionary steps or side-tracking maneuvers," which are not only unhelpful but lead in the wrong direction altogether.

III

> Our longing for the beautiful seems to shine from it, situating us within its disclosed world that has its own new and surprising urges. This longing is not in our possession, but conveys an *inexhaustible depth* of object that belongs to it while exceeding it. The *depth* is a signal within the form of the object whose intention embraces us. Given this simultaneous objectivity and subjectivity of the experience of the beautiful, we can say that to see . . . the beautiful is *to see the invisible in the visible.*
>
> *John Milbank*[7]

The strange and dreamy history of aesthetics has resulted in several anomalies and perplexities that have important bearings on the depth, the interiority, and that alluring radiance of nature that has traditionally been called beauty. Of these, two are striking and immediate, while two more are best understood as derivative from them.

Most perplexing of all is that reflections on beauty in aesthetics have themselves been so few in number. "To say of beauty that it is relatively unexplored," observes Mary Mothersill, "is just to observe that many great philosophers treat it in a perfunctory manner or not at all. . . . The popular conception—that there is a vast [philosophical] literature [on beauty], many theories, as many as there are theorists—is false."[8] Why this longstanding neglect?

A second, striking perplexity is that even when beauty does get discussed, what is itself less puzzling (the beauty of art) is taken as the central problematic, while what is more puzzling, and indeed most wonderful (the beauty of nature), is ignored altogether or taken as marginal. That is, it is the beauty of *art* that is the customary locus and norm of the discussion, even though the occurrence of beauty in *nature* is far more remarkable philosophically. For it seems no more exceptional or puzzling that works of art could be beautiful than that propositions could be true or that human actions could be good: In each case, it seems relatively unproblematic that each could be respectively beautiful, true, or good simply because they were each intended to be that way, made to possess those characteristics. But that a sunset or a tree or a bird can be found to be beautiful should appear to us just as remarkable as someone claiming, without making any qualifications, that one of these was good or that it was true. Surely it is just as exceptional to say that a given tree is beautiful as to say that it is good or true—that it exhibits of its own accord one of the sorts of things which, more than being profoundly human, are what is profoundly constitutive of our humanity itself—the sorts of things that Plato maintains are prerequisites for our having turned out to be human in the first place.

Art is *made to be beautiful*, while it is hardly self-evident to philosophy that the intentionality of an artisan lies behind the beauty of nature.[9] Nevertheless, our modern ordinary ways of speaking and thinking easily accommodate something's being beautiful just on its own, and in its own right, whereas to say of a cliff-face or a wildflower that it was simply good, and even less so that it was true, would puzzle most listeners. And accordingly, philosophical theories have found this easy to accommodate as well. As is so often the case, it is puzzling how often what is truly puzzling is not taken to be puzzling at all. The perplexity, and wonder, of natural beauty—that it should be at all—has hardly been addressed by philosophers.

IV

> In the first place, here [upon this very twig laden with orange blossoms] is a prodigality of beauty; and what harm do they do by existing? And is man not a being capable of Beauty even as of Hunger and Thirst? And if the latter be fit objects of a final cause, why then not the former?
>
> *Samuel Taylor Coleridge,* Anima Poetae[10]

Leo Tolstoy, who in the end reduces aesthetics to morality, nonetheless usefully distinguishes between two philosophical conceptions of beauty. "The first," he explains, "is that beauty is something having an independent existence (existing in itself), that it is one of the manifestations of the absolutely Perfect, of the Idea, of the Spirit, of Will, or of God; the other is that beauty is a kind of pleasure received by us not having personal advantage for its object." The first, the "objective-mystical definition," Tolstoy finds characteristic of German Idealism with the second, "subjective" definition, characteristic of British aesthetics. In the first, beauty is "something mystical or metaphysical," what Muir called "beauty beyond thought," while in the second it is simply "a special kind of pleasure."[11] At the risk of oversimplification, we could add that the first view traces its lineage to Plato and sees the locus of beauty in what is extraordinary, exceptional, transcendent, perhaps sublime. Its exemplars are natural beauty rather than art, moral beauty (especially when usual human limits are transcended, as with heroes and saints), and the visionary beauty described by mystics. The second finds its predecessor not in Plato but in Aristotle, seeing the locus of beauty in human affectivity and imagination, and taking art as its exemplar. When it does consider nature, it valorizes the harmonious, the balanced and proportional (i.e., the pastoral).

It is, then, the predominance (especially in modernity) of the second kind of aesthetic—the kind that sees beauty as subjective, as mental, as a kind of pleasure—that helps account for the two earlier perplexities. For if beauty is simply the pleasure we get from certain impressions, then it *is* something quite unremarkable, either in art or in nature, a matter of no more than our psycho-physical constitution.

But is beauty thus described really the beauty we seek and encounter, either in nature or in art? Or is it instead a tamed and bloodless remnant of the rich, dynamic beauty that, in our most fortunate moments, addresses us? Is the sublimity of a thunderstorm, or a mountain range, or indeed of "starry skies above" really nothing more than my cozy sentiment of moral autonomy, as Kant believed? Beauty itself is not just puzzling, but wonderful, something transcendent and theophanous. And most wonderful of all is that in the natural world, we discover a beauty that we did not create and that does not inhere in our own minds. What Tolstoy calls subjective theories of beauty, rather than helping us understand these mysteries, cover them up instead. Moreover, the predominance of such views accounts for the second set of anomalies as well, those that are specific to environmental philosophy proper.

The third perplexity can be stated as follows: given the predominant role of natural beauty in understanding and articulating why nature is important—both in native, uninstructed sensibilities and in our finest nature writers—it is strange that aesthetic considerations have played such a marginal role in environmental philosophy. For the guiding vision of American nature writing is already familiar to the simplest camper or hiker or hunter: We should protect and preserve nature because it is splendid, singular, beautiful. If the writer finds a missionary task, it is either to enlarge and expand our native aesthetic sensibilities (to include reptiles or arctic tundra or inhospitable swamplands) or else to revive and refresh our encrusted or sedimented memory of nature's beauty—to recall us to our own best perceptions and our own best selves, to help us see better, farther, more clearly that which all of us have already surveyed. It is never aimed at generating this sensibility in the first place, for as Plato insists, it could not be imparted if it were not present already.

Why, then, do aesthetic considerations get overlooked in environmental philosophy? At one level, it is probably because in postmodern, pluralistic societies it is widely believed that aesthetics is utterly individualistic and entirely subjective, not just in the academic sense of inhering in the subject, but in the popular sense of being completely relative to the whims of different individuals. Some people, it would follow, prefer the aesthetics of forests while others prefer the aesthetics of shopping malls—preferences that would neutralize each other, and render aesthetic arguments for environmental protection ineffectual. But at a deeper level, not just subjectivistic theories of beauty, but even those ontologically grounded theories of beauty that are current today, fail to account for what is most important about the beauty of nature. That is, they fail to articulate an aesthetic of the holy in nature—that phenomenon of depth or integrity or eternity or transcendence that is always grasped, however tacitly or adequately, within the beauty of nature, and that constitutes the living heart of that beauty itself.

Once this is clarified, we can address yet a fourth perplexity, one that also entails a failure of aesthetics. This perplexity arises from environmental philosophy's longstanding neglect of some of the most urgent causes for concern, as well as some of those that are more subtle, yet for that reason more insidious:

> The radical threat to the integrity of our atmosphere that results from global climate change
>
> The far-reaching assault upon life itself through bio-engineering, and the related danger of genetic pollution

The global suppression of silence though noise pollution, and the blocking of the starry skies themselves through global air and light pollution

The extinction of species, seen not as depletions of the "gene pool," but as a sealing over of nodal apertures of eternity

Here, too, simple sensibilities are ahead of philosophical reflection. Just as those lovers of nature who are philosophically unsophisticated proceed unapologetically from nature's beauty, so they are often just as unequivocal in their revulsion at these onslaughts as modes of sacrilege, cosmic blasphemies, raging unholiness. A consideration of the holiness that is present within natural beauty, will show that these expressions are not hyperbole at all, but rather precise articulations of the danger, of what we risk to lose.

Through briefly considering two German philosophers (Nietzsche and Heidegger) juxtaposed with two more thinkers (Solovyov and Florensky) representative of the powerful, but largely unexplored, field of Russian religious aesthetics, the remainder of this essay will outline an understanding of the holiness of nature as encountered within the beauty of nature. Such an aesthetic will necessarily be of the ontological, or if it is preferred, mystical sort. And it will conclude, as already noted above, with suggestions about how this understanding can help explicate some of the most grievous, and grievously neglected, assaults upon nature that are at work in the world today.

V

For John Muir, the high mountains were revelatory—as much an aesthetic revelation as a religious epiphany—merging the sublimity of stony heights and the smiling beauty of gentle meadows with an intense, pervasive sense of the eternal. In articulating this, Muir sets the benchmark for nature-writing in America, which always seems to either anticipate his sensibilities (as with Thoreau and Bartram) or to echo them (as with virtually all of his successors). But Muir also offers us a key for reading certain passages in Nietzsche, who found inspiration not while trekking the peaks of the High Sierra, but hiking in the Upper Engadin Alps and in the high mountain cliffs above Rapallo and Portofino. It could, indeed, as easily have been Muir as Nietzsche who reported that his guiding revelation was received "6000 feet above man and time."[12] Like Moses and

Elijah, like his own Zarathustra, like John Muir, Nietzsche goes up into the high country, a mile above humanity, to receive his revelation—which unfolds a mile above time as well, for it is a revelation of eternity. Nietzsche himself characterizes this revealing as "inspiration," "rapture," and "revelation": "the concept of revelation—in the sense that suddenly, with indescribable certainty and subtlety, something becomes visible, audible, something that shakes one down to the last depths and throws one down—that merely describes the facts. One hears, one does not seek; one accepts, one does not ask who gives . . ."[13]

The entire course of Nietzsche's life and thought ascends to this vision—a trajectory that rarely departs far from the aesthetic considerations with which it began in *The Birth of Tragedy*, a work that sought to embrace the ecstatic wildness of the Dionysian, along with the measure and harmony of the Apollonian. Nietzsche's Zarathustra, articulating this highest vision, inaugurates his teaching with a call to "be true to the earth," loyal to the visible and audible and palpable. But tellingly parallel to mystics and religious visionaries of many traditions, he concludes that the highest fidelity lies in seeing the invisible *within* the visible, the eternal within the temporal, i.e., the seal of eternal recurrence imprinted within the very heart of the transitory. This is not an abstract conclusion, but a creative vision, the high point of Nietzsche's aesthetics. It is the highest creative vision—a vision from the heights—since it creates a new world, a new and redeemed worldredeemed from time and its having-been, and hence redeemed from the vengeful project of creating a competing, parallel world. The teaching of eternal recurrence offers, as Nietzsche had maintained throughout his works, a certain kind of aesthetic vision as redemption. Moreover, this is no longer an aesthetic filter through which we view the world in order to avoid perishing of truth, an imaginative nosegay that would spare aestheticized nostrils, but a transforming of that reality itself. Yet in what way?

Perhaps not in the same way that Nietzsche thought. For Nietzsche attempts here to revive the ancient project of *noēsis* in a manner that is nonetheless faithful to *aisthēsis*—to see the eternal and invisible *within* the sensible and temporal—by means of affirmation alone, through "saying yes" to the eternal repetition of everything just as it is, in its banality and its everyday superficiality, granting each moment the depth of eternity. Yet endless repetition is not eternity at all, but sempiternity, endless duration, Hegel's "bad infinite," perhaps even the endless etcetera of madness. The thought of eternal recurrence is first articulated in *The Gay Science*, where it is whispered that everything, even "this spider in the moonlight," will recur endlessly. This image eerily evokes an exchange in Dostoevsky's *Crime and Punishment* (whose influence may perhaps be seen in the "Pale Criminal" chapter of *Zarathustra*) in which even the ax murderer Raskolnikov is shocked at Svidrigailov's proposal that eternity should consist of nothing more than the endlessness of a room full of spiders and desiccated webs. The comparison reminds us that the endless recurrence of the selfsame spider on the selfsame windowsill adds nothing to the original spider, but rather erodes it. Nor if, logically, it must be the very same spider and room and the exact same dreariness and banality that is affirmed, could affirmative gestures possibly alter anything, interjecting eternity or anything else into the situation, lest the identity of the same be forfeited?[14]

Ironically, it is Zarathustra's vision not of spiders, but of his own animals—of his snake and his eagle—that tacitly moves toward accomplishing what eternal recurrence fails to do. The snake and eagle are not mere "symbols" in an allegorical sense, but real animals who are at the same time symbols in the root sense of *syn-holon* (i.e., they bring-together-into-one the visible and the invisible). They are more than just a reptile and a bird because Zarathustra sees more in them, sees them better, grasps their significance more deeply than others, even his followers. This is why he takes joy in them, for as Zarathustra teaches, all joy loves the depth of eternity, "deep, deep eternity." But can Nietzsche's philosophy of eternal return account for this deeper seeing without either reducing it an aesthetic overlay—an enhancement designed to make life more interesting, more attractive—or else trying to forcibly infuse the visible with the depth of eternity through a sheer determination of will? How can it be understood as genuine *noēsis*, as authentic seeing?

VI

A more far-reaching solution than Nietzsche's aesthetic of will and affirmation can be found in the work of the Russian philosopher Vladimir Solovyov, friend and confidant of Dostoevsky, and Nietzsche's junior by nine years. Like Nietzsche, Solovyov experienced a series of mystical visions of the eternal made manifest in the visible. (Niezsche's inspirations took place in Switzerland and

Italy, Solovyov's in his native Russia, in England, and in Egypt near the pyramids of Giza.) Parallel to Nietzsche's call to "be true to the earth," Solovyov called for a recognition of our "solidarity with mother-earth," and even warned (in 1891) that the result of "our false view that nature is lifeless matter and a soulless machine" has been that "earthly nature, as though offended by this double untruth, has refused [even] to feed mankind."[15] If Nietzsche envisioned the world's redemption in the eternal return of the same, Solovyov foresaw the redemption of nature in his vision of Sophia, the Divine Wisdom linking God and nature. Solovyov draws upon the Wisdom literature of the Old Testament, Plato's *Timaeus*, Jakob Boehme, and above all on the liturgical and iconographic traditions of the Eastern Church to understand Sophia as "the highest and all-embracing form of the living soul of nature and of the universe, united to God from all eternity and in the temporal process attaining union with Him and uniting to Him all that is."[16]

Byzantine Christianity—the understanding of which has eluded the West for a millennium and a half—has always seen salvation as a cosmic event, a redemption of nature and all creation as well as humanity; seen the Incarnation as a glorification of nature and of matter itself; and seen the final resurrection of the end times as a perfection and transformation of the earthly. (Hence its greatest cathedral, "Hagia Sophia"—the Church of the Holy Wisdom of God—displays the world's first great architectural dome, to present the dome of heaven itself come down to earth, the merger of the visible and the invisible.) Solovyov employs this understanding of Sophia as the basis for a universal aesthetic—one of the few in the history of philosophy that comprehends the beauty of art through the beauty of nature, rather than vice versa.

> Of the two kinds of beautiful appearances—nature and art—we will take first that which is wider in extent, simpler in content and prior to the other in the order of time. The aesthetics of nature will give us the necessary basis for a philosophy of art.[17]

Here, too, the imagery of mountains is at play. Evoking the Transfiguration of Christ on Mount Tabor, Solovyov sees the beauty of nature as its own transfiguration, its luminosity revealing both its integrity and its fulfillment. Like Plato, Solovyov sees beauty as the shining forth of the invisible through the visible, but he does not see the visible as merely a ladder to be kicked away once it is surmounted. Nor, as with Hegel, is the sensuous appearance subsumed within the ideal, leaving behind its particular and transitory character as an outworn husk. Rather, this infusion of the temporal with the eternal is by that fact a revealing of the inner beauty of the visible itself, its perfection. "Nature itself," he argues, "is not indifferent to beauty," but moves toward it, inclines toward it. Before art transfigures nature, nature transfigures itself as its inmost tendency. "Beautiful animals," he maintains, as we recall Zarathustra's eagle and serpent, "express the intensity of vital motives united in a complex whole and sufficiently balanced to admit free play of the forces of life."[18] And the more this inner reality, this manifesting of their own spiritual content, shines through, the more beautiful are the animals. And the more perfect the manifestation of the eternal. And the more completely the creature is its own self. Three different descriptions of a single movement. The beauty of art, then, is simply the anticipation and elaboration of this tendency which exists already in nature. The burning of the bush on Mount Sinai, Solovyov maintains, expresses this illumination from within by the eternity of which each being is already an image. Finally, Solovyov maintains a continuity between *erōs* and *agapē*, and argues that ultimate redemptiveness comes through that love which apprehends and affirms the individual other (be it another person, or a fallen leaf) in its own integral and unrepeatable character—in the eternity of its own idea.

VII

It is possible, however, to move beyond this ultimately Platonistic sense of the *eternal*—which captivated Nietzsche, albeit in a negative manner, as much as it did Solovyov—toward the specific phenomenon of the *holy* in nature. This will also allow us to see why it is that nature itself—nature as such and as a whole, and not simply natural individuals—is best understood under the determination of the holy. It will, in addition, bring to light a historical element that is indispensable for understanding certain larger, global issues that environmental philosophy has left largely unaddressed, as well as suggest the unavoidability of a dimension of personal engagement.

No contemporary philosopher has rivaled Heidegger's expositions of the holy in nature in its aesthetic dimensions, many of them interpreting the poetry of Hölderlin. "The holy," Heidegger states, is the very "word of Hölderlin's poetry"—"the claim under which his poetic activity stands."[19] Moreover, "the holy" is the means by which Hölderlin attempts to "think" that which he calls "nature": not the theoretical nature of the natural sciences, but "divinely beautiful nature," nature as *physis* in the ancient Greek sense, as the shining-forth of unity within the oppositions of rain and sunlight, night and darkness, heaven and earth, humanity and divinities. "Beauty is the pure shining of the unconcealment of the whole infinite relation, together with its center."[20] Hölderlin's writings, Heidegger suggests, comprise a poetic commentary on the interplay between these three elements (nature, beauty, and the holy) as well as on the poet's relation to them. For Hölderlin, the holy (*das Heilige*) is the element within which alone it is possible to find the traces of *divinity*. The traces of the *holy*, in turn, are to be found within the hale and wholesome (*das Heil*). But so bereft is our own age of these integral and healing elements (*das Heilen*) that we are surrounded instead with the unwholesome, the un-whole, the un-healthy—with a world in which we are not only unable to find the traces (*Spuren*) of divinity that we are unable to find the traces of the whole and healing (*das Heil*) themselves.

With the un-whole, the un-wholesome, the unsalutary (*das Heillos*) as our environment, we increasingly find ourselves surrounded not by the holy but the unholy (*das Unheil*). This epochal, but inconspicuous, occurrence has come about, Heidegger argues, through modern technology, which has effected a series of eclipses, occlusions, or closures within the sphere of all that is salutary and wholesome—air and sunlight, the stars overhead, primordial silence, the very food we eat. Indeed, as Heidegger writes in his "Letter on Humanism," "perhaps what is distinctive about this world epoch consists in the occlusion or closure (*Verschlossenheit*) of the dimension of the salutary or wholesome (*des Heilen*). Perhaps that is the sole malignancy, the singular unholiness [*Unheil*]."[21] Heidegger's conclusion is that a massive unholiness is enveloping our world today, as a result of the global character of technological mode of concealing that he calls *Gestell*. The very element of nature in its holiness (i.e., divinely beautiful nature)—nature as the beautiful unity of the infinite relation—is undergoing a global closure, an obfuscation of its beauty and integrity, yet one that is inconspicuous and entirely compatible with the preservation of "scenic landscapes." This, in turn, is due to the localized occlusions, themselves global in their effects, of the salutary, the integral, the wholesome. What are some examples of these specific closures? Looking ahead to the last section, we can suggest that one is the closure through which—even in the most remote parts of the earth, now susceptible to overhead air traffic—silence has become at best, in the few times and places it can be found, a temporary abatement of mechanical and electronic noise. Another is the unsalutary and unholy closure through which the very climate that mediates the natural environment to us is no longer a gift of the earth and sky, but a stale by-product of technological processes. Heidegger argues, then, that the poet in such a world must seek not the traces of the holy in nature—this would be too much to expect in dark times—but merely the traces of the salutary that would give us access to the traces of the holy.

VIII

> Objectivity does exist. It is God's creation. To live and feel together with all creation, not with the creation that man has corrupted but with the creation that came out of the hands of its Creator; to see in this creation another, higher nature; through the crust of sin, to feel the *core* of God's *creation*. . . . How is one to understand this holy, this beautiful aspect of creation?
>
> *Pavel Florensky,* The Pillar and Ground of Truth[22]

The similarities between Martin Heidegger and Pavel Florensky are as impressive and instructive as the differences. Both were born in the same decade (the1880s), and each went on to become the leading philosophical intellect of his respective country (Germany and Russia). Both were raised in rural areas near the mountains (the Black Forest, and the Trans-Caucasus), and both sought to retrieve their respective traditions in ways that, ironically, often led them into regions of the cultural avant-garde. Both confronted modern totalitarianism, but while Heidegger collaborated with the leadership of Nazi Germany, Florensky openly resisted Bolshevik tyranny, and for this was eventually murdered in the gulags of Stalinist Russia in 1934—the same year Heidegger resigned his rectorship and abandoned his attempt to "lead the leaders." Florensky—perhaps the twentieth century's most notable polymath, with impressive publications in philosophy, theology, linguistics, art history, mathematics, science, and engineering—was silenced in mid-career and his writings banned, so his work is only now being discovered in Russia, in Germany and Italy, and to a lesser extent in America.

Florensky understands what Heidegger calls "the holy" by means of the ancient concept of Sophia or Divine Wisdom, originally revived by Solovyov and drawing upon the same cosmic spirituality of Eastern Christianity. Thinking nature as "creation," Florensky argues (parallel to Heidegger) that "Sophia is all-integral creation and not merely *all* creation. Sophia is the Great Root by which creation goes into the intra-Trinitarian life [of the Godhead] and through which it receives Life Eternal from the One Source of Life. Sophia is the original nature of creation."[23] At the same time, "Sophia represents the spirituality of creation, its holiness, purity, and immaculateness, i.e. its beauty." Sophia *is* Beauty, the "essential beauty in all of creation" that shows itself to "the contemplation of creation in its unity."[24] So far, then, Heidegger and Florensky agree: Nature/creation seen in its salutary or integral character shows a kind of higher unity that is beauty of the highest kind and the essential manifestation of holiness/divine wisdom. They are even parallel in their distinctions of two kinds of thinking—calculative versus meditative thinking in Heidegger, and discursive versus noetic understanding in Florensky—one of which can approach nature in its beauty and holiness, with the other obstructing the way.

But pursuing the historical and collective dimensions of the obstruction, Heidegger discovers modern technology, whereas regarding the blockage spiritually through his exposition of "person," Florensky sees not too much order, but rather disorder: "an encrustation of the heart," the isolation of "insane selfhood, [i.e., of a mode of] selfhood that has lost its mind"; like Solovyov, Florensky sees the problem in a failure of love.[25] Citing Nicholas Stithatos in the *Philokalia*, he maintains that "the nature of things changes according to the inner nature of the soul" (200), and in his biography of his own spiritual master, the Elder Isidore, Florensky portrays a simple man whose outwardly impoverished life is rich with contemplation of the beauty and holiness of nature, while exhibiting a remarkable compassion toward its most humble creatures.[26] The socio-cultural encrustation, then, through which modern humanity has cut itself off from the deeper beauty and holiness of creation, and the shallow obsession with discursive rationality and its potential for mastery and control, are both collective forms of perennial temptations, and in Western Europe and its sphere of influence both have become culturally ensconced and even valorized. Nor is this to be overcome through thoughtfully awaiting the "word" of a great poet, but only by divesting ourselves of the shallow self-enclosure that isolates us both from one another and from the holiness and beauty of nature (i.e., through learning to love). Florensky, then, provides an important complement and corrective to Heidegger: the *closure* of the salutary and integral, and of the holy and beautiful in nature, is correlative to the *en-closure* of the human heart in a deranged isolation that has traditionally been called "sin" and "iniquity," and which can be overcome not by thinking and poetizing alone, but through what has traditionally been called a "change of heart."

IX

And did those Feet in ancient time
Walk upon England's mountains green?
And was the holy Lamb of God
In England's pleasant pastures seen?
And did the Countenance Divine
Shine down upon those clouded hills;
And was Jerusalem builded here
Among those dark Satanic mills?
Bring me my bow of burning gold,
Bring me my arrows of desire:
Bring me my spear, O clouds unfold,
Bring me my chariot of fire.
I shall not cease from mental fight,
Nor shall my sword sleep in my hand,
Till we have built Jerusalem
In England's green and pleasant land!
William Blake, "Jerusalem"

At a loss for words, a consumer advocate, recently interviewed on a TV news show, maintained that health risks aside, the public simply did not want to consume food from cloned animals, due to what she called "the 'yuk!' factor." She was, of course, politely ridiculed by the interviewers for her efforts. At a more sophisticated level, the popular author Bill McKibben has been derided for his broadly articulated claim, in *The End of Nature,* that global climate change subverts and effaces the sacred dimension of nature. One of the things that philosophers can offer today is a lexicon for articulating the dangers we face: dangers such as artificial, engineered food that is not merely a temporary trigger of irrational squeamishness, but that is repulsive on a deeply aesthetic, a profoundly theological, a radically ontological level, for it is food that is unwholesome, unsalutary, *unholy*. It has severed any perceptible continuity with—broken with any last traces of—the earth itself. Dangers such as these raise a prospect of the loss of nature's very naturalness, of that inner, intact dimension—of that *depth* dimension—that renders to it a most important kind of sanctity.

We need not, of course, rely exclusively, or even primarily, on philosophers and deep ecologists to tell us of this threat. We sense it already, and we know it though our great poets: not just Hölderlin, but Emerson and Thoreau, Muir and Burroughs, Dillard and Lopez and Snyder. Their articulations of the salutary and whole in nature—and of its integral beauty—are so overwhelmingly consistent that they have led a hostile critic (Joyce Carol Oates) to strangely complain that nature "inspires a painfully limited set of responses in 'nature writers'—REVERENCE, AWE, PEITY, MYSTICAL ONENESS."[27] (It is hard not to wonder: does the author really think these are somehow not nearly enough, i.e., that they are much too "limited"?)

Yet philosophers can help clarify these matters, and pursue them into unexpected areas, by showing connections between issues such as wilderness and species preservation (long ago sheltered within the province of environmental discourse) and seemingly disparate issues such as global warming, bioengineering, light and sound pollution, urbanization, and a host of others. In each case, there is need for a natural aesthetic more radical than the pastoral or even the sublime: need for an aesthetics of the hale and integral versus the disrupted and fragmented and degraded—and ultimately an aesthetics of the holy and the unholy. Such an aesthetic of nature can show, philosophically, that evaluations such as these are not subjective nor are they of secondary importance. Rather, they concern some of the most important parameters of a world that would count as a world at all: that those many apertures of integral beauty—windows upon the eternal and holy—not be closed off in the radical manner characteristic of modern technology. What does it mean that few people today have seen a starry sky, or experienced real silence—or that we no longer trust the food we eat to be a natural offering, something offered up by the self-emergence of *physis* itself?

X

> And for all this, nature is never spent;
> There lives the dearest freshness *deep down things*.
>
> *Gerard Manley Hopkins, "God's Grandeur"*

In 1877, at the crest of the industrial revolution, with the landscape of Blake's "dark Satanic mills" close by, Gerard Manley Hopkins writes of this holy beauty of nature: "The world is charged with the grandeur of God. / It will flame out like shining from shook foil"—that is, as he later explains, shine out in the way that "shaken gold-foil gives off broad glares like sheet lightening." Yet immediately the poet adds: "Generations have trod, have trod, have trod; / all is smeared with trade; bleared, smeared with toil; / and wears man's scent and shares man's smell: the soil / is bare now, nor can foot feel, being shod." But again, there is a reversal: "And for all this, nature is never spent; / there lives the dearest freshness deep down things." In this "dearest freshness deep down things," the poet articulates in a phrase what this essay has tried to elucidate at some length. A century and a quarter later, we ask: Can the soil become so hardened, and the human heart itself become so encrusted in its thirst for resonance in the world, that the life-giving depth of this "freshness deep down things" becomes all but inaccessible, and for this reason, socially and culturally and politically irrelevant? Perhaps it is, above all, this depth that is at risk in all assaults upon nature. Perhaps it is this, more than anything, that is ultimately at stake in our environmental concerns, and thus cannot be understood through either scientific or ethical categories alone, but only by means of aesthetico-theological considerations such as those pursued here. Perhaps it is, as well, the depth within which deep ecology finds its own truest roots.

CHAPTER

2

Nature's Other Side

The Demise of Nature and the Phenomenology of Givenness

The idea of a universe that is self-subsistent—standing entirely on its own, fully operational and intelligible, independent from anything outside itself—is both odd and modern. In the course of human experience, it is an extraordinary concept, defying the shared wisdom of virtually all peoples, almost everywhere outside of Western Europe and its sphere of influence. And even within this orbit it is distinctively, and in its fully articulated configuration, exclusively modern. It is today commonly taken to be one of the great achievements of Western culture, paving the way for modern science and an enlightened understanding of nature in general. Environmentalists, too, pay tribute to this concept, feeling that it somehow honors nature to regard it as self-subsistent. But this chapter shall maintain that it is instead one of the most disastrous ideas of modernity, inimical to any salutary relation between humanity and the natural environment. The concept of nature as autonomous, as self-subsistent or self-contained, has for several centuries deeply confounded modern thinking about nature. More recently, it has blinded environmental philosophy both to the unitary character of the problems that it faces, as well as to the direction to which it needs to look for answers. It is a conception of nature, and stands for a specific relation to nature, that is inherently one-sided, for the cosmos that it discloses possesses only one side.

Superficiality

Disentangling the visible from the invisible made it "inhuman" in our minds, by reducing it to mere matter. At the same time, this made it appear capable of being wholly adapted to humans, malleable in every aspect and open to unlimited appropriation.

Marcel Gauchet, The Disenchantment of the World: A Political History of Religion[1]

An outside alone, pure exteriority, is only a surface. It is a plane, a superficies; it is sheer extension. It cannot present a face, for there is no inside to face out. And because it has no inside, it is not strictly speaking an outside at all. It is mere superficiality, unrelenting superficiality, and it is all-inclusive superficiality: not merely surface, but triviality, a surd surface, just there and nothing more. Lacking an interior, the surface relinquishes no face; it cannot face us, but is merely present. There is, indeed, a kind of alterity here, but it is dead and inanimate. Offering up neither invitation nor response to our gaze, it stops the gaze, deflects it, reflects it. In its faceless reflectivity, it sparkles, and even dazzles. The sheer surface, just there, *vorhanden,* is in its dazzling superficiality, a mirror.

To the seeker, the mute surface with nothing behind it, and nothing within it, seems meaningless. The saying "What you see is what you get" seems to him not just glib but cruel. Nature now becomes absurd, the Kingdom of Sisyphus. But it is also possible to cast down the lantern and abandon the search, to enjoy the marketplace and delight in the mirroring surface, to glibly embrace slogans proclaiming that "there is no depth," and "there is no other side." Are not the play of light on the surface, the sparkling reflections it casts, meaning enough? Nature now becomes semiotic, postmodern, a carnival fun-house of mirror after mirror. But before this, it must first have become a surface, an exterior.

Nature, for modernity, is a second-order exteriority, a faceless surface not in any of its parts, but as a whole. Nature as a whole becomes an externality, with nothing behind it and nothing within it. But in relation to what is this externality exterior? It is most immediately exterior to the knowing subject, standing outside it as pure object, extension confronting cognition, *res extensa* standing over against *res cogitans*—an externality that is not spatial but ontological. Nature is substance, self-subsistence that is exterior to knowing, standing opaquely on its own over against the self-transparency of consciousness. And this remains true for Hegel, for whom nature precisely *is* exteriority, just as much as it is true for Descartes and the philosophy of the *cogito.* Hegel indeed purports to free nature from its self-subsiding externality, but he proceeds in the wrong direction, sublating nature *as* nature, rendering it as artifact, as art, as *Begriff*, as *speculum* for the speculative consciousness, as absolute self-reflection of the knowing subject, as pure self-seeing in a mirror whose glassy surface allows the gaze to feel it is not superficial at all, but rather penetrating and deep. As was already seen by Marx, who wished to restore, and indeed radicalize, this externality of nature—precisely to find in our actual appropriation of it the true mirror of our humanity—Hegel had never really left it behind at all. In Hegel, as well as Marx, just as for Descartes, nature is exteriority, self-subsistence, and mirror.

For Hegel—and in large part for German Idealism, British Romanticism, and American Transcendentalism—nature is a mirror through which spirit can gaze upon its own face in the belief that its gaze is not superficial, but deep. And for that parallel tradition, from Bacon to Marx, that seeks to exploit nature directly, to appropriate it in actuality, nature is a mirror reflective of what is specifically human: For this gaze, unfolding possibilities gleam back from nature like the glitter of gold for the gaze of *conquistadores.* Nature here is a mirror as well, the surface for a "mirror of production," through which capitalist and communist alike, and in equal measure, can find a faithful reflection of human possibilities.[2]

The surface upon which the gaze rests, that forms a mirror for the gaze and allows it to see its own reflection, is what Jean-Luc Marion has called an "idol." But it is, he argues, the gaze that forms the idol, not the idol that shapes the gaze. The idol itself "results from the gaze that aims at it."[3] It is thus a production, our own construct, even as it stands over against us in an externality that somehow passes for a givenness. Marion has articulated a phenomenology of the onto-theological consciousness. And he has shown us that the onto-theological gaze, the characteristic vision of modernity, is the gaze of an idolater. For Marion, the god of onto-theology—the god to whom Heidegger has said we can neither sing nor pray, and before whom we can never fall down on our knees in awe—is just an idol.[4] But if Marion is right about this, then neither is Heidegger entirely correct: that is, we can indeed make prostrations before such a god, but in doing so we bow down precisely as idolaters.

What, then, is the cherished projection screen, the jeweled speculum, that mirrors back the idolatrous gaze unto itself? What is it other than nature? And what likeness does this luminous superficies of nature reflect back to the modern gaze? My possibilities of wealth, for the productionist. Mathematical rigor, for the scientist.

My deepest self, for the romanticist. Community, for the environmentalist. Always the reflected likeness of the idolater's gaze.[5]

But has nature no interiority of its own—has it only an exterior? Has it no "integrity" of its own, as Kohak has called it, that would confound the idolatrous gaze?[6] Is nature, as Gadamer has said, nothing more than our own possibilities of using it.[7] Has nature no face of its own to confront, and convert, the idolatrous gaze?

Interiority

> We are far from ascribing any sort of idolatry to Adam, who understood God's language, now effaced from our memory and, as we have observed before, vainly sought by our poets. Adam blessed not the matter of trees and stones, but the spirituality of Creation, the single Body. . . .
>
> *O. V. de L. Milosz,* The Arcana[8]

A face requires an inside. A face is inside-out—*is* the inside facing out. From what had once been a surface alone, not yet even an exterior, now an interiority faces us. What faces us has an inside, and what has an inside is alive.

If nature faces us, addresses us, responds to us in a meaningful way, does this mean that nature is alive—not just certain "organic" entities in nature, but nature as such? Surely the ancient Greeks believed this—as in one way or another, do pre-modern peoples generally. David Abram has articulated this sensibility for—this orientation toward and apprehension of—an inner life of nature drawing upon the notion of "participation mystique" in Levy-Bruhl, as well as Merleau-Ponty's phenomenology of perception.[9] Max Scheler, nearly a century ago, also discussed this sensibility at length in his phenomenology of sympathy, under the concept of *Einsfühlung* toward nature, the "feeling of unity" with nature, the emotional identification with it that is necessarily oriented toward all nature as living, as organic, as somehow ensouled. Scheler describes this sensibility as "an immediate and non-inferential leap into the living *heart of things,*" one that it is not possible to hold toward "dead matter," but which presupposes an apprehension of "*all* natural phenomena . . . as the undivided total life of a single world-organism," and thus as the "outward expression of the inner life thus imparted." It is this all-pervasive inner dimension, and not some imagined personality traits, nor a metaphysical pan-psychism, that allows

Martin Buber to say "*Du*" to a fragment of mica.[10] And parallel to Abram, Scheler concludes emphatically that this "organic conception of the world in one or another of its thousand forms, has to this day held practically undisputed sway over all but the western portion of humanity."[11]

For the Renaissance West, this interiority is *natura naturans*, nature naturing. For the Byzantine East, it is *Hagia Sophia*, the Holy Wisdom of God: "the *spirituality* of creation, its holiness, purity, and immaculateness, i.e. its beauty."[12] For the ancient Greeks, the world-soul of Plato and Plotinus—as "divine as a changing thing can be"—was the high-water mark of this sensibility, outpacing the Greek poetic consciousness through its apprehension of nature as a whole, and not simply certain extraordinary phenomena, as being ensouled. Nor does this understanding attribute the features of life to each of its parts, for this would commit the fallacy of division. It is nature as a whole that possesses this interiority, this animation; it is not the specific character of every being (stones and bubbles and flames, such as the Jains are said to believe) to be empirically alive. Moreover, this sense of a unified life of nature that gets articulated in the metaphysics of the world-soul counters and checks the centrifugal tendency toward a congealing of cosmic life into nodal points. That is, it impedes its condensation into particular entities that would be, for example, the specific life of the thunderbolt itself as well as, at the same time, the daimonic being who hurls the thunderbolt. In the ancient world, the principal defectors from affirming this holistic "inner life" of cosmic nature were the Greek atomists, who saw nature as self-subsistent and self-contained, a mechanistic system of lifeless atoms with no interior and no second side—a one-sided understanding that was soundly rejected by both Plato and Aristotle, as well as by their successors, and that remained marginal until its revival and eventual predominance in the modern age.[13] Plotinus, however, saw clearly that for there to be an inside, there must at the same time be another side, a second side turned away from our gaze—turned penultimately toward what Neoplatonic metaphysics called the *Nous*, and turned finally toward the One.

But far less celebrated at the time (and perhaps even more marginal today) was another poetics operating at the other, eastern end of the Mediterranean, one that had already grasped with much greater clarity the connection between (1) the sense of a unified life of nature and (2) the sensibility for a second side, another side, a third dimension of nature. In the Psalter, that classic anthology of

Hebrew poetry and song, is to be found an aesthetic appreciation of the inner life—and indeed, inner voice—of nature *as a whole* that is almost entirely lacking in classical Greek poetry. To the environmentally inclined reader not deafened by monodic paraphrases of Lynn White Jr., the Psalms can present themselves as containing the first, and perhaps the greatest, collection of nature poetry in the literature of humanity. In these songs, the inner life of nature as a whole emerges with surpassing power and beauty: In one litany after another, the entire range of cosmic nature—sun and moon, rocks and hills and mountains, fierce winds and lightening and crashing waves, diminutive serpents and skittering rock badgers—all, in concert, sing out in praise, rise up in song, skip and leap and dance and whirl in joy and anticipation, tremble in awe. This is by no means a performance for the sake of humankind, as if for their ruler, but rather these voices and songs and rumblings and this dancing are consistently portrayed as utterly overshadowing humankind with a sublimity that was not to reappear in the art and poetry of the European West until the nineteenth century. That is, nature presents a face here, expresses an inner life, only because it is at the same time disclosed as being turned radically and ecstatically toward a distance unto which all the resonance of that life is directed, and from which that life is itself derived. It is, moreover, toward that distance above all that these "psalms" or "songs of praise" (Hebrew: *Tehillim*) are meant to resound; only consequently are they intended for human ears.[14]

Givenness

We created a second Nature in the image of the first
So as not to believe that we live in Paradise.

Czeslaw Milosz, "Advice"[15]

In the perceptual realm, things face us only because at the same time they turn away. One side is turned toward us—let us name it, in relation to ourselves, as the front side. It is more than just a surface only because it has as its counterpart an other side, a side that is turned away, a back side, a concealed side. And even if we walk around the thing, say a tree or a hill, it keeps one side turned away even as we try to see it from all sides. This is, of course, precisely what allows it to have an inside, to be more than a single surface, a superficies, a strangely hallucinatory Möbius strip.

To have an inside is at the same time, and necessarily, to have an other side, a side turned away, withheld. But if it is nature as a whole that is at issue here—that can be disclosed either as self-subsistent and as superficies, or else as having an interiority or inner life—it would need to be nature as a whole that was to be encountered as having an other side. What, then, would this mean? And what would be the mode of encounter, the sensibility to which it could be disclosed? Historically, it is clearly the mytho-poetic sensibility—against which modernity, from its inception, has waged a war without quarter—that has encountered nature in this manner. Such a consciousness—the pre-modern and extra-modern norm for humanity, and as Scheler argues, still "the primordial and 'most natural' way of regarding nature"—sees the cosmos as a marvel, a wonder, as uncanny, as mysterious, but never as just there, as self-contained and self-subsistent.[16] As is suggested by both David Abram and Annie Dillard, the consciousness that is charmed in this way by the natural world in which it finds itself exhibits something of the delighted response elicited by sleight-of-hand magic: "Now you see it! Now you don't!"[17] That is, such an awareness apprehends nature as a whole as being wondrously given, as being bestowed in a marvelous way. For this sensibility, nature possesses a kind of causality so extraordinary that it is perhaps best not to call it causality at all, but rather a radical givenness, a bestowal, or—to employ a mytho-poetic word, an aesthetico-theological word that onto-theology has tried to appropriate—creation.

The other side of nature is the side that allows it to be more than a mere *Gegen-stand*—an ob-ject standing counter to a sub-ject—and more than our own production.[18] It is the back side, the other side, that allows it to have an inside. The other side is the side that we sense but do not see, that is turned away from us toward that distance from whence it is rendered, out of which it is bestowed, and to which—retaining a certain fidelity—it inclines. For nature-as-a-whole to possess this other side, then, would entail that it be encountered as possessing as-a-whole a radical givenness, as issuing from a strange kind of bestowal unlike any other. This would not be a causal inference committing, as Hume maintained, a fallacy of composition. The mytho-poetic sensibility is not discursive or inferential but noetic, and rather than reasoning from the efficient causality within the parts of nature to that of the whole, it overlooks this efficient causality altogether in favor of a radically different kind of givenness. That is, the so-called cosmological arguments first formulated at the dawn of modernity, and which may be taken as early cairns of onto-theology, were new symptoms of the unfolding eclipse of mytho-poetic sensibilities that made modernity—and self-subsiding nature—possible in the first place.

This eclipse is already evident in the formulation of an even earlier kind of proof, Anselm's ontological argument. Here, as the Latin West first begins to assert its theological acumen, Anselm repeatedly cites the verse from the Thirteenth Psalm, which states: "The fool has said in his heart that there is no God." Anselm discusses this passage as if the "fool" were foolish because of mistaken inferences, leading him to think, erroneously, that God could exist in the mind but not in reality. But the tradition from which Anselm was, by the process of this very reflection, cutting himself off (along with his successors) began not with the mind at all, but with real nature, lived nature—and proceeded not discursively, but poetically. Prominent glosses on this verse, certainly well known to Anselm, can be found already in the Wisdom of Solomon and the Epistle to the Romans.[19] And both state clearly and emphatically that the fool is foolish not at all because of deficient reasoning ability or defective inferences, but rather because of a hardly conceivable failure of spiritual vision: the inability, or rather refusal, to see the radical givenness of the cosmos in the beauty which everywhere surrounds us in nature—the same beauty evoked throughout the Psalms as almost irresistibly eliciting a chorus of praise to its creator.

For pre-modern sensibilities—and this means for most people in most times and places—it took an exceptional blindness, so extraordinary that willfulness had to be assumed, in order to fail to see this other-sidedness of nature, to miss the uncanniness of the strange beauty that surrounds us, to overlook the invisible that is ubiquitously, and alluringly, and engagingly manifest in the visible. For the modern world, however, almost the converse is true. In a cosmos that is experienced as self-subsistent and self-enclosed—external to the onto-theological gaze, and as seen from within that gaze, external to the source of givenness itself, since such a cosmos seems not to have been given at all—in nature as it is encountered in modernity, the exercise of this noetic vision is itself both extraordinary and marginal, to be accomplished mostly either by monks and mystics on faraway mountains, or by poets who in increasing numbers end up going mad in the process, even as neither are acknowledged to be undertaking something serious anyway. But many of the poets who do not go mad—poets such as Thoreau, Muir, Snyder, Berry,

Dillard, and Lopez—are those who seek out both the other side, as well as the inside, of a nature from which they refuse to remain external. It is to them, and to the ancient mytho-poetic religious sensibilities that they struggle to perpetuate, that we need to turn if we are to adequately understand the genuine scope, and the appropriate modalities, of environmental philosophy.

Foremost among these modalities would be a phenomenology of givenness addressing, among other things, how this givenness is itself given. This is not, however, entirely lacking—nor should this fact be surprising if we understand phenomenology as the attempt to wrest from modernity the noetic vision that modern thought itself had occluded. Husserlian phenomenology persistently tries to articulate the very givenness (*Gegebenheit*) of phenomena, as well as the prerequisites both for encountering phenomena in their pure givenness, and for encountering that process of givenness itself. Heideggerian phenomenology traces modalities and degrees of givenness beginning with the derivative, and deficient, mode of encountering entities as just there, as self-subsistent, as merely "given" rather than in their givenness, that is, as *vorhanden*. But our encounter with beings, and especially natural entities such as trees and streams, unfolds more primordially when they emerge in their bearing upon our concerns, as they are implicated in our meaningful dealings with the world, that is, as they are handy or *zuhanden*. The progression here is not just from less to more primary, but from self-subsistence to an emergence and givenness that Heidegger will eventually articulate as *physis*, self-emergence and self-unfolding. Later yet, and particularly in his essays on the poet Hölderlin, he shows how this more primary mode of emergence is yet more complex, enjoining the poetic word, and most especially corresponding to the saying of "the holy," within which nature can show itself most fully, and whose very name becomes for Hölderlin a higher invocation for nature as a whole.[20] Finally, Heidegger situates the possibility for this emergence within an ultimate Event, an *Ereignis*, which he explicitly understands as a Giving, a *Geben*.

These strands are taken up suggestively in the work of Jean-Luc Marion, yet in a way that is strangely a-cosmic and an-aesthetic. Marion's work on the phenomenon of giving—and which he defends against Derrida's critique of the possibility of giving—lends itself at many points to a phenomenology of *nature's* givenness, of its other-sidedness. For example, in contrast to the idolatrous gaze of onto-theology, that is, of modern thought and sensibility, Marion posits an iconic vision that allows—much as do the icons of Eastern Orthodoxy—the invisible to be encountered in the visible, and thus allows for a face to emerge for the first time. Meanwhile, citing Hölderlin's poem "In Lovely Blueness . . . ," he shows how the givenness that makes possible this iconic vision—this seeing of a rootedness of the visible in what is not visible—requires a distance that withholds itself even as it gives, precisely so that the giving can take place: a distance that remains manifest, in the words of the poet, but "only like the sky." Such a giving, he argues, at the same time entails a "call" to the one to whom it is given. Unlike the idolatrous gaze, which provides itself with its own object, iconic seeing must be responsive.[21]

Perhaps even this brief characterization may suggest how a phenomenology of nature's givenness, if developed farther in these directions, could offer a philosophical articulation of those exemplary encounters with nature that are voiced poetically by our finest nature writers, and that with surprising accord show how the only nature that we will ever find worth preserving is one that has an inside and thus another side, a relation to something entirely "other" that allows us to find it not just valuable, but venerable. Such a phenomenology would warrant, I maintain, the highest priority for environmental philosophy to arrive at a solid self-understanding.

But if we must first find nature venerable—as possessing an other side, as issuing from a "luminous darkness"—before we are really likely to preserve it, from what must nature be preserved?[22] And how should we understand the nature that needs to be protected and preserved?

Demise

> So spreads the appearance that everything encountered has being only to the extent that it is a human construct.
>
> *Martin Heidegger,* "The Question of Technology"[23]

What has been severed from its source of life, and thereby deprived of its interiority, is by that fact dead. A one-sided cosmos is a lifeless idol, the dumb product of our own hands—yet it somehow, and paradoxically, maintains itself as an artifact whose icy silence, even amidst the ensuing "breath of empty space," does not inhibit our vain and absent-minded worship of it.[24]

The tropes of an assault upon nature, and of a correspondingly threatened demise of nature, have been at play since the dawn of modernity. While Francis Bacon employed the language of an assault upon nature (he spoke of hounding it, subduing it, constraining it, vexing it, conquering it, squeezing and molding it, and dissecting it) his contemporary John Donne was writing:

> The Sunne is lost, *and th' earth*, and no mans wit
> Can well direct him, where to look for it.[25]

This sense of cosmic displacement and demise reaches a high point in the twentieth century with, to give only a few examples: T. S. Eliot envisioning a literally God-forsaken land where spring does not arrive; Rachel Carson (thirty years later) warning of a devastation of nature so thorough that the spring really would be silent; and in the seventies and eighties, two different books (one in French, and another in English) called *The Death of Nature*, leading in 1989 to Bill McKibben's similarly titled book, *The End of Nature*. To what extent are such images merely figurative?

For McKibben, at least, the threat of an end to nature is not figurative at all. Rather, it impends menacingly in the imminent loss of the very givenness (and hence, naturalness) of nature that has been resulting from human influence upon global climate patterns, and by means of which the elemental conditions of weather on earth—the sun on our skin and the wind on our face—come to be seen as themselves human products, or perhaps better, "by-products." It has quite sensibly been suggested that McKibben is somewhat naive in his belief in the pre-modern purity of nature, as well as rigid in demanding such a pristine character for nature to be itself. It is—yes, of course—a matter of degree. But at what point does nature get wrested from its givenness to so many degrees, and in so many ways, that it is no longer nature at all, until at last it really is nothing more than *vorhanden*, "just there," a lovely corpse in urgent need of "preservation"?[26] And why, if this is the genuine danger, has environmental philosophy failed to recognize the radical character of the threat, as well as the compass of its variations?

Why, for example, has the escalating threat of global climate change not so far been a serious and sustained issue for environmental philosophy? And why not, as well, the siege upon nature that is being carried out by bioengineering? Or by the genetic modification of the foods we eat? Or the offensive that is at work in the contempt, underwriting genetic cloning, toward the inestimable, unrepeatable, breathtaking *given*-ness that announces new life? Why is the assault upon nature's integrity that is perpetrated by factory farming seen almost exclusively within the narrow rubric of animal rights? Why, indeed, is the attack upon perhaps one-fifth of all human pregnancies in America today not seen as an assault upon nature—our nature—in its very givenness, and thus as a most troubling issue not just for bioethics and religious morality, but for environmental philosophy? And why have environmental philosophers overlooked the links between these rather conspicuous onslaughts upon the givenness—the very naturalness—of nature, and those more commonly thematized assaults associated with species depletion (generic givenness) and the shrinkage of wilderness (aggregate givenness)? Are not all of these attacks different fronts of that single, great, protracted war upon nature that *is* modernity? And doesn't this accelerating assault upon nature as a whole, this ontological assault upon its givenness, provide us at last, and far better than the conceptual framework of "ethics," with the only coherent and unified subject matter for environmental philosophy, as well as a pivotal insight into what ought to be its overriding goal: the rescue of nature from its own demise—i.e., from its becoming so thoroughly and so comprehensively uprooted from its inherent givenness that it actually *becomes* what since the dawn of modernity it has been *understood to be*: a self-subsistent, self-contained system?

My conclusion, then, is simply that the *ontological demise*, which has long been underway, has been masked and concealed by the *aesthetico-theological demise* in modern thought and sensibility, under whose influence we have learned to encounter nature as if it were, in fact, already dead. And to the extent that environmental philosophy bases itself exclusively, or even primarily, upon the discursive rationality of the sciences—which by their very methodology deliver nature as a self-subsistent, self-enclosed system—rather than upon the mytho-poetic vision, which approaches nature iconically, as issuing from a distance and an invisibility that enchants it and animates it, environmental philosophy is complicitous in this ontological demise, precisely through inadvertently concealing it. It follows, then, that only by recognizing nature's beauty and integrity as entailing a disclosure of its own other-sidedness—that is, by reorienting itself away from the modernist and scientific vision, in the direction of a noetic, and thus an aesthetic and religious vision—will environmental philosophy identify both the extent and the coherence of its own subject matter. And only in this way is it likely to be helpful in the task of restoring to nature not only conceptually, but actually, its own interiority, along with the givenness that is proper to it. For without this givenness and this interiority it is no longer nature at all, but just another mirror—a glassy and shallow and lifeless sea—reflecting that which for its own part is one-sided as well—reflecting back that which is human, all-too-human.

CHAPTER

3

Layers of Nature in Thomas Traherne and John Muir

Numinous Beauty, Onto-theology, and the Polyphony of Tradition

I

No Western philosopher has provided a richer context for addressing environmental issues than Martin Heidegger. According to Heidegger, the ancient (and perhaps future) experience of nature as *physis* or "self-emergence," as the site of "the arrival of the gods" and thus as "divinely beautiful," has been supplanted in modernity by the experience of nature as object (*Gegenstand*) and ultimately as resource (*Bestand*). This has, moreover, come about through the predominance of what he calls onto-theology, a development inaugurated within ancient Greek philosophy, but which becomes fully constituted only within Christendom and its post-Christian bequest.

In his earlier writings, heavily influenced by his reading of the early writings of Luther, Heidegger argued that an original, more "existential" level of Christianity had been layered over by the onto-theological metaphysics deriving from Greek philosophy, especially as it culminates in Aristotle's claim that "first philosophy" must be both ontological and theological. Heidegger understood early Christianity as concerning itself with the experience of certain modes of "facticity" and temporality, leading him to conclude (along with many Protestant theologians of the time) that ontological and cosmic aspects of patristic thought and spirituality represented not legitimate attempts to comprehend the implications of this new religion for the natural world, but rather pagan intrusions and corruptions of its basic mode of "life experience." Thus, not only in his 1920–21 lectures on the phenomenology of religious life, but also in many of his other courses and writings of the twenties (most notably, in *Being and Time*), Heidegger attempted to excavate and retrieve these putatively "pure," lived structures of existence from what he would later term their onto-theological overlay, preparing a staging area for phenomenological insight through an approach that he called in German, *Destruktion*—"destruction," but less in the sense of annihilation (*Zerstörung*) than of the German *Abbau*: "de-structuring" or "de-construction," but also (parallel to Husserl's notion of "de-sedimentation") perhaps "de-layering." If we add to this Heideggerean notion, and its incalculable influence upon at least four generations of philosophers and theologians, and their reading of texts, the enormous influence in English-speaking countries of Derridean "deconstruction"—which follows from Heidegger's approach through certain seemingly minor adjustments, that nevertheless result in a radically dissimilar project—it has become one of the hallmark methodological concepts of the postmodern era.

Yet as was first advanced by Pöggeler in 1963—and as has more recently been exhaustively demonstrated by Van Buren, Crowe, and a number of others, the original inspiration for this philosophical tactic of *Destruktion* was Martin Luther, whose theological views are now widely acknowledged as having had a profound, far-reaching, and enduring influence upon Heidegger's work. Because this influence has been perhaps the most important factor in what might with some justice be called Heidegger's refusal of immanence—making anything like a "noetics of nature" a highly questionable enterprise for him—and because its enduring legacy in Heidegger's thought has lent his *Seinsgeschichte* a univocity that becomes increasingly narrow, Heidegger's early relation to Luther must be clarified in some detail.

II

In 1916, Heidegger completed the final version of his *Habilitationsschrift*, or qualifying dissertation, on the medieval philosopher John Duns Scotus. Here he maintained that rather than standing in a relation of opposition (*Gegensatz*) to one another—embodying such false dichotomies as rationalism and irrationalism—medieval scholasticism and medieval mysticism were not merely complementary, but mutually supportive of one another, dual aspects of a single medieval experience of God and world: "For the medieval worldview [*Weltanschauung*], scholasticism and mysticism belong together in an essential manner."[1] Thus, Heidegger (still identifying with Catholicism and Neo-scholasticism) argues that the explication of mysticism will show that scholastic concepts are not mere abstractions, but are deeply rooted in life experience. Yet these claims remain merely programmatic in this work, with mysticism itself being mentioned in only two passages, one of them merely stating the urgency of working-through phenomenologically both the "mystical" and the specifically "ascetic" writings of the period, and the other promising to complete a "philosophical explication and valuation" of the mysticism of Meister Eckhart.[2] What would have surely seemed like an obvious preparation for accomplishing these goals, then, was Heidegger's course on "The Philosophical Foundations of Medieval Mysticism," announced for the Winter semester of 1919–20. But the course was cancelled in late August of 1919, after Heidegger had spent only a few weeks in formal preparation, citing insufficient time to prepare, even though he had been deeply immersed in the literature of Western medieval mysticism since at least 1910.[3] The course on medieval mysticism never appears, and only scattered references to Eckhart follow in Heidegger's published work—Caputo counts only seven of them subsequent to 1930.[4] Instead, Heidegger delivers two lecture courses on themes that now seem to have superceded the earlier course on mysticism: one course on St. Paul, and the second on St. Augustine, both seeking to show how authentic Christianity is rooted in a mode of experience that is quite different from Heidegger's earlier understanding of the mystical. Finally, from this time on, his references to mysticism are now almost entirely disparaging.[5] What has happened in the three years between 1916 and 1919?

It seems clear that Heidegger underwent a crisis in his own religious faith during this time, probably well underway by 1917.[6] And perhaps its most obvious manifestation was his absorption in the writings of the early Martin Luther and the profound influence of Luther not only on his understanding of Christianity, but upon his view of the task of philosophy. Under Luther's tutelage, Heidegger was now able to find something surpassing the *basis in life experience* for scholastic thought that he had once sought in mysticism. In the writings of St. Paul, he now found a radically *new modality of life itself*, one that was bound up with a new experience of temporality, and that he would soon be calling "facticity" (*Faktizität*).

In the Heidelberg Disputation of 1518, in Heidegger's view "the most pointed formulation of Luther's position in his early period," Luther had posited two competing theologies, mutually exclusive options that to ancient Christians would have surely seemed like the terms of an entirely, and needlessly, false dilemma.[7] First, Luther vituperates what he regards as a retrograde "theology of glory" (*theologia gloriae*), which he believes retreats to the pagan contemplation of God in nature; which seeks contemplative "rest" and "peace" in the divine presence; and which inevitably results in what Luther regards as pride and a sense of self-fulfillment. In the words of Heidegger's enthusiastic glosses upon theses 21 and 22, "the *theologus gloriae* who aesthetically takes delight in the wonders of the world" gains a "wisdom that sees what is invisible of God from His works, [and thus that] inflates us, blinds us, and hardens our hearts."[8]

On the other hand, Luther valorizes what he calls a "theology of the cross" (*theologia crucis*), which proceeds not from the experience of God in nature, as traditional wisdom had always done, but rather begins with a *destruction* (*Destrucktion*) of that contemplative experience and of what Luther believed was its natural result: a state of being that has become arrogant and blind, as he believed was the case with Greek metaphysics and above all with his scholastic adversaries—however remote the *theōria* of the latter might now seem from the contemplation of nature in figures like Traherne or Muir, or indeed like Dionysios and Maximos. Or for that matter, like the Psalmist, who cannot sing enough praises of the cosmic glory of the Creator. But for Luther, all those who would seek the glory of God in the created order are really just seeking their own glory, not that of God. Remarkably, Luther even sees the "theology of glory" as a kind of cruelty. For since the Fall, which the young Luther regards as having radically disfigured the created order, nature itself is now crying, lamenting, and mourning over its own corruption. And "is he not a madman" who would take contemplative enjoyment in "someone who is crying and lamenting?"[9]

The *theologia crucis*, proceeding from the *Destrucktion*, is now free to seek God not in what is beautiful, but in what is offensive and repellent; not in what is glorious and magnificent, but in what is weak—i.e., not in the created order as it has always seemed to manifest its Creator, but in the cross and suffering where God is "hidden." That is, God chooses to reveal himself in suffering, weakness, humiliation, and crucifixion—things that are entirely contrary to the divine nature. But this means that God is not visible in nature at all, contrary to the ancient understanding of creation in both Jewish and Christian traditions; rather, God becomes *visible* only in the very elements that would serve to hide and obscure the divine being itself. That is, even within the radical immanence of the

Incarnation itself, God reveals Himself as *deus absconditus*. The invisible God's self-manifestation in visible creation is a revealing that is in fact a concealing, a non-presence, an absence. For Heidegger, then, this *theologia crucis* would count as nothing less than a radical destruction of what he would come to see as the metaphysics of presence.

But what about the innumerable texts in the Old Testament (especially in the Wisdom Books) leading to the opposite conclusion—for example, in the Wisdom of Solomon (13:5), which insists that "through the greatness and beauty of created things, their Creator is correspondingly seen"? Or what of Jesus commending his followers to look at the glory (*doxa*) of wildflowers in order to see something important about God's care for creation? Or especially problematic for Luther and Heidegger, who both see St. Paul as embodying the *theologia crucis*, what to do with Paul's own gloss (in Romans 1:20) upon the Wisdom of Solomon passage just cited: "For since the creation of the world [*kosmos*] the invisible things of God are noetically seen through the things that are made, both His eternal power and His divinity"? One hopes that the seemingly ingenuous circularity of Heidegger's response to this difficulty, as transcribed in Becker's protocol, was in fact more cogently elucidated in the delivered lecture itself: "Only *Luther* really understood this passage [Romans 1:19] for the first time."[10]

For Heidegger, then, all this comes as a revelation. He becomes convinced that Luther has rediscovered original Christianity, Christianity in its pure and primal form, and therefore that traditional articulations and elaborations need to be dismantled, or more literally destroyed. Dismantling this "debris," he is able to read St. Paul freshly, and the first course on the Philosophy of Religion (WS 1920 –21)—i.e., the first offering to serve as a substitute for his planned mysticism course—focuses upon just such a reading of St. Paul's Epistles to the Thessalonians and the Galatians. Here, what Heidegger had once sought in mysticism is now found in the "factical life experience" described by St. Paul, and which for Heidegger becomes the model not only for his *Daseinsanalytic* in *Being and Time*, but for his later critique of metaphysics. One of the most important texts in his exegesis is taken from Paul's First Letter to the Thessalonians (5:1–3):

> But as to the times and the seasons [*tōn chronōn kai tōn kairōn*], brethren, ye have no need that I write unto you. For you yourselves know perfectly that the Day of the Lord so

> cometh as a thief in the night. For when they are saying, "peace and security" [*eirene kai asphaleia*] then sudden destruction cometh upon them as travail upon a woman with child, and they shall not escape.

Luther's *Destruktion* of the *theologia gloriae* would seem to have little connection to this passage, but for Heidegger it is evident and decisive. All *theōria* or contemplation of the *kosmos*, and thus Greek metaphysics as a whole, is nothing more than a quest for the very "peace and security" against which the Apostle warns the Thessalonians. It is an attempt to find "rest" in what is present, in the traditional sense of *parousia* or "presence," i.e., the eternal glimpsed within the temporal. But here, Heidegger argues, one finds instead a new sense of *parousia*: nothing present, but the futural coming of the "Day of the Lord," which will arrive suddenly, unexpectedly, like "a thief in the night." Thus, we must not seek rest in any kind of divine presence, take any delight in immanence, but rather seize upon distress, weakness, and anxiety as revealing the divine absence, and hence the "hidden" God of the *theologia crucis*: "Only when he is weak, when he withstands the anguish of his life, can [St. Paul, and thus the authentic New Testament Christian] enter into a close connection with God . . . Not mystical absorption and special exertion; rather withstanding the weakness of life is decisive."[11]

Moreover, this "constant insecurity" is "not coincidental; rather it is necessary."[12] Any attempt at certainty regarding the coming *parousia*, any attempt at "speculating cognitively about it," is nothing less than to fail "the time of testing" associated with the Antichrist, rather than holding firm in "the highest anguish" and "absolute distress."[13] Only from within this uncertainty and anguish, this radical not-having of God, can the genuine Christian exercise the anticipatory hope of the *parousia*, an eschatological temporality of the futural, which will in turn become central in all of Heidegger's thought: a *deferral* of presence, and ultimately a skeptical *refusal* of anything that might offer a fulfilled present, of any transcendence become manifest and immanent—a resistance to *any theophany or transfiguration*—as the temptation of the Antichrist, and therefore as belonging to "the danger" or *die Gefahr*, to use Heidegger's later terminology. Here, Heidegger believes, he has found the authentic, "primal Christianity" that almost immediately gets covered over—even in apostolic times, through the infusion of

Greek metaphysical speculation, and even in the later works of Martin Luther himself. As Vattimo has noted, along with others, "The importance of the '*Einleitung*' [i.e., Heidegger's lecture course, "Introduction to The Phenomenology of Religion"] is not only that it anticipates clearly and effectively many themes of *Being and Time* [but that these themes are] linked here *essentially* to a meditation on Christian experience, along with the idea of metaphysics as the forgetting of being, which Heidegger would develop later."[14] Thus, the inner destiny of Western thought and experience is put into play here with the Hellenization of "primal Christianity," and its reversion to the quest for constant presence against which Heidegger will struggle throughout his career as a philosopher: Here the plot-summary is laid out with surprising comprehensiveness, with only a few brief "eruptions" of exceptional individuals (Luther, Pascal, Kierkegaard) who re-discover (usually only briefly and tentatively) the authentic temporality that Heidegger has found in a passage from St. Paul, under the tutelage of the great Reformer.

But can such a simplistic approach to two thousand years of Western thought and experience be anything more than a historical mold that is hardly less rigid and formalized than Hegel's view of history as the advance of freedom? Isn't this really only a sophisticated variant of Harnack's views, popular at the time, of the "Hellenization of Christendom," or worse yet, a late triumph of the proto-fundamentalist and heretical Tertullian, who excoriated any complicity between Athens and Jerusalem? Are there not at every point other voices, and other strata at work in a slow and complex plate-tectonics of religious culture? Can such an overarching understanding of Christian history really be based upon a reading, however insightful, of a few passages from St. Paul's Epistles, especially given the lack of evidence that Heidegger had any serious familiarity with patristic sources prior to Augustine? Can Christian mysticism be so easily rejected, on the basis of what is, for the most part, a study of only three figures (Meister Eckhart, Bernard of Clairvaux, and Teresa of Ávila), who are themselves relatively exceptional when viewed from within the wider Christian tradition, especially as it unfolded in the Byzantine East, of which Heidegger appears to have had virtually no familiarity at all?

Yet even Heidegger's determination to restore a sense of factical existence and its authentic temporality—along with its later modalities—is not itself univocal, since it possesses layers that are easily overlooked. For example, Heidegger had begun reading Dostoevsky at the same time he had discovered medieval mysticism, and he kept a photograph of Dostoevsky on his writing desk throughout his working career. Could the Eastern voice of the great Russian thinker have failed to have woven itself, however subtly, into Heidegger's meta-narrative? And despite his emphatic and consistent rejection of anything resembling a noetics of nature as an inauthentic reversion to the "peace and security" of presence, and of *theōria* as mere gazing, we must also consider Heidegger's sustained study of certain poets (all of whom seem to have qualified for their standing by writing from within a situation of distress and anguish) as exceptional visionaries of what such a fulfilled noetics of nature might at least someday be. Does this not count as something like a *poetics of nature*? And perhaps through this poetics, Heidegger himself can be seen as outlining what, with some anachronism, might be regarded as a *transcendental* noetics of nature—i.e., as developing the "conditions of the possibility" for what a fulfilled experience of the invisible manifest within the visible *would be*—perhaps as a component of some coming *Ereignis*, perhaps as the blessing of a *letzte Gott* in some unforeseeable future.

Within a few more years, Heidegger has broken with traditional modes of Christianity altogether, and by 1928 he is not only testily accepting the rubric of "atheism," but is now suggesting (as he will continue to do) that the refusal of what he calls "ontic faith" by the "genuine metaphysician" is actually "more religious than the usual faithful, than the members of a 'church' or even than the 'theologians' of every confession," suggesting that the real "godlessness" lies with the presumptive believers.[15] Increasingly, Heidegger looks to the experience of earth and gods (*die Göttlichen*) in archaic, preclassical Greece, to orient him toward a more salutary relation between humanity and nature, even if it is one for which we can do no more than "prepare the readiness of awaiting"—as he puts this in the *Beitrage*—for the final god (*die letzte Gott*).[16] And what Heidegger now explicitly terms onto-theology is called into question no longer for undermining the soul in its inner quest for authenticity and salvation, but for subjecting being itself (and above all the earth) to an imperative of mastery and control.

It can be argued, however, that the Western tradition, both as it derives from Athens as well as from Jerusalem, is far less monolithic than Heidegger's *Seinsgeschichte* would have it, especially concerning the triadic relation between nature, humanity, and divinity. Both these traditions, ancient Greece and ancient Judaism, rather than maintaining a monolithic discourse concerning humanity and nature, offer instead a polyphony of discourses that are interwoven and layered through three millennia, and that have been drawn upon in diverse ways by a variety of figures. The religious and literary traditions of both ancient Judaism (especially in the Psalms and Wisdom literature) and traditional Christianity (especially as it is preserved in Eastern Patristic and Hesychast practice and writings) are oriented far more toward a poetic encounter with nature and cosmos—approaching them above all in their beauty—than Heidegger was prepared to recognize. Even Platonism, especially as it developed in the twelve hundred years after Plato in Alexandria and Cappadocia and Constantinople—historic centers of the Byzantine East—was often far more oriented toward the lived or "existential" preconditions for encountering the beauty of nature, than Heidegger acknowledges.

The remainder of this chapter, then, will argue that certain of these interwoven layers of Western thought and sensibility (indeed, even as they have been progressively obscured and written-over in Scholasticism, the Reformation, and the Enlightenment, and overlain most seductively by Heidegger's own narratives) have quietly made possible a vision of nature, largely aesthetic in character, that has not only occasionally resurfaced from time to time in Western Europe—suggestively in figures such as St. Francis of Assisi, and more explicitly in Thomas Traherne and John Muir—but also formed the operative, driving force for the environmental movement in Western thought. Although many other modern writers (such as Goethe, Hölderlin, Emerson, Thoreau, Burroughs, Snyder, Dillard, Lopez, and Berry) could easily be recruited for support, this thesis will here be argued largely with reference to the writings of Thomas Traherne, a seventeenth century English poet and Anglican priest, and John Muir, Scottish-American writer and naturalist of the late nineteenth and early twentieth centuries, and more appropriately than anyone else, richly deserving of the title of founder of modern environmentalism.

III

Few Christian writers would be more disconcerting to the dour understanding of Christianity embraced by the early Luther, and after him by Heidegger in the 1920s, than Thomas Traherne, for whom contemplative joy in the glory of God, as it is experienced in creation, is not just permissible, but normative for the authentic Christian life—requisite not for speculative metaphysicians and "mystics," but for the everyday believer. Or perhaps, indeed, we should read him as enjoining each believer to become himself a mystic.

In his *Centuries of Meditation*, as well as in his *Christian Ethicks*, Traherne challenges the prevailing view of nature in the seventeenth century, which he associates especially with Thomas Hobbes. For example, contrary to the overarching Hobbesean themes of self-preservation and possessive individualism, Traherne argues that the only self we would really *wish to preserve* is not some monadic self, isolated from relationships, but the self in love—the self along *with* its love, and thus along with *what* it loves: the "great mistake in that arrogant *Leviathan* [is to] imprison our love to our selves," overlooking the things that the self loves, and that we must hence also love along with the self, if we are to truly love the self at all. "These are the ends of Self-preservation," Traherne insists. "And it is impossible to love ourselves without loving these." Self-love itself delivers us *to* the love of these things, to the desire of them, to the enjoyment of them.[17] And what is lovable above all, what is most to be enjoyed, is "the world"—i.e., in Traherne's lexicon, not the world of convention and worldliness, but quite the opposite: "nature" in all its God-mirroring, God-exhibiting, God-sharing beauty, or as Muir will put it, its "Godful" glory. Traherne writes:

> You never Enjoy the world aright till you see how a Sand Exhibiteth the Wisdom and Power of God . . . Your Enjoyment of the World is never right till evry Morning you awake in Heaven: see your self in your fathers Palace: and look upon the Skies and the Earth and the Air, as Celestial Joys: having such a Reverend Esteem of all, as if you were among the Angels . . . You never Enjoy the World aright, till the Sea it self floweth in your Veins, till you are Clothed with the Heavens, and Crowned with the Stars: and perceiv your self to be the Sole heir of the whole World: and more then so, becaus Men are in it who are evry one Sole Heirs, as well as you.

He adds:

> Yet further, you never Enjoy the World aright, till you so lov the Beauty of Enjoying it, that you are Covetous and Earnest to

> Persuade others to Enjoy it. . . . The World is a Mirror of infinit Beauty, yet no Man sees it. It is a Temple of Majesty, yet no Man regards it. It is a Region of Light and Peace, did not Men Disquiet it. It is the Paradice of God. It is more to Man since he is faln, than it was before. It is the Place of Angels and the Gate of Heaven.[18]

Remarkably, Traherne has passed rather effortlessly here from self-interest and the desire for enjoyment to an evangelical beneficence that seeks to share with others this enjoyment of the world. Of some interest, too, is his resemblance here to John Locke, who sought to mollify the bellicose effects of Hobbesian selfishness through what he felt was a cooperative motive inherent to private property, and the need of peace to improve and preserve it. Yet here, Traherne seeks to go much farther, exhorting each reader to see himself as the "sole heir" of *all* the world's beauty, while at the same time further showing that to fully enjoy this inheritance entails sharing it with others and exhorting them to see *themselves* as sole heirs as well, and so on, making his prescribed practice truly a "region of Peace."

Traherne sees this "enjoyment" of the world as that which should be our most characteristic relation to it, taking "true enjoyment" (as opposed to mere possession) as a celebration of the world as it presents itself to the senses, yet one that at the same time slides effortlessly into what has sometimes been called a "nature mysticism," a joyful contemplation of divine beauty in every twig and leaf and indeed in every grain of sand. And the presence of God is what is most enjoyable about this beauty, and our ability to discern it is what is most wonderful about us.

But this insistence upon divine immanence is by no means pantheistic, as if taking the inherence of God in the world as substantial, but rather insists that the divine presence in nature is a gift of grace, whereby the world shines with light, transfigured like the figure of Christ on Mt. Tabor, transparent to the divine energies. "That God should give us soe Divine a Power!" he exclaims. "To Transfigure all Things and be Delighted."[19] And this power of transfiguring enjoyment is simply the ability—too often hardly exercised at all—to love. "Lov is the true Means by which the World is Enjoyed. Our Lov to others, and Others Lov to us. We ought therefore abov all Things to get acquainted with Lov: for Lov is the root and Foundation of Nature. . . . The very End for which GOD made the World was that He might Manifest His Lov. Unless therefore we can be satisfied with his Lov so manifested we can never be satisfied. There are many Glorious Excellencies in the Material World, but without Lov they are all Abortiv."[20] There is, then, an engaging circularity here. God is love, and manifests this love in the beauty of the world, which can therefore be apprehended only by the same kind of love through which the world was created—one that wants to share its enjoyment of manifested love with others.

Finally, the ability to love—and thus, too, the ability to enjoy the holy beauty of nature—both depend upon humility, which Traherne argues "makes men capable of all felicity." "All deep apprehensions" and indeed all our spiritual powers are "contained in humility. There they grow deep and serious and infinite; there they become vigourous and strong; there they are made substantial and eternal. All the powers of the soul are employed, extended and made perfect in this depth of abysses."[21] This path or "motion from vice to virtue" is one that will recur in various modes throughout the present study, for it is the ancient path to *theōria* and *noēsis* by means of *katharsis*.[22] It is a way that is understood not only by all great mystics (of whom I believe Traherne is one) but I am convinced by many of the greatest poets as well (among whom, once again, I believe that Traherne deserves inclusion). Traherne even articulates something close to the patristic view of humanity as constituting a priesthood of creation. As put by S. Sandbank, "The human soul, knowing the lower in terms of the higher world, collects it, purifies it, and lifts it up to universality, thus restoring the unity of Being itself."[23] By means of this consecration, the visible world is united with the visible, earth with heaven, in what Traherne does not hesitate to call "Marrying the Creator and his Creatures together."[24]

Traherne stands here as a link and precursor not only to Blake (who urges us to cleanse and purify "the doors of perception," that we might see how "every thing that lives is holy") but also to Emerson and Thoreau and a whole succession of American nature writers, many of whom (like Muir) felt a constant need to purify the spirit through leaving civilization and inhabiting the wilderness—as did the great prophets and visionaries of earlier times. But with Traherne, it is particularly clear how deeply steeped is this sensibility in the literary universes of both the Psalms and the great Wisdom Books and Prophetic Books of the Old Testament. And yet at the same time we find here along with the *philokalia* or love of beauty that is a hallmark of all Neo-Platonism, a legacy that was much more explicitly developed in the Byzantine East, where (contrary to

Heidegger's exegesis) "*theōria*" was not an onto-theological exercise in "correct vision" and objectification, but rather was linked to "the book of nature" itself as "*theōria physikē*," a seeing of the visible in the invisible, of what Heidegger himself calls "divinely beautiful nature," and that seems strikingly to further the program that Annie Dillard lays out much later in her exuberant essay on "seeing."[25] As Traherne puts this in verse:

> Sure Man was born to meditat on things,
> And to contemplat the Eternal Springs
> Of God and Nature, Glory, Bliss, and Pleasure;
> That Life and Love might be his chiefest Treasure . . .
> D'ye ask me What? It was with Cleerer Eys
> To see all Creatures full of Deities.[26]

IV

In the nature writing of John Muir, the mystical contemplation of divinity as revealed within nature, and the poetic types and figures of the Psalms, are layered and interwoven into an unassuming idiom of American vernacular. Being raised initially within the Calvinist world of the Scottish Lowlands, with his later youth spent on a Wisconsin farm in a harshly religious household, Muir had memorized most of the Bible by his teens. He went on, in turn, to become a religious visionary himself, as the scholarship concerning Muir is beginning to realize, and he is gradually coming to be seen more as transforming the Christianity of his birth than as rejecting it.[27] In fact, he held fast to many of the theological views with which his youthful thinking had been infused, albeit in a much transfigured state, and more important, he came to embrace even more emphatically the sensibilities and ecstatic language of the Psalms, the rustically hortatory character of the Wisdom Books, and the enigmatic vision of the natural world prevalent in the New Testament parables. Indeed, Muir very much became (like his father had been) an evangelist and missionary, but an evangelist of a new sort, exhorting his audience to experience and celebrate God as manifest in nature and especially in wilderness. So much does the outcome differ from both the earth-denying "Platonism for the people" criticized by Nietzsche, as well as the onto-theology depicted by Heidegger, that Muir has often been portrayed as tacitly embracing an essentially pantheistic or pagan or even Taoist outlook. (As we will see in the case of Dostoevsky and others in this book, when the deepest and most ancient affinities of the Christian view of nature become manifest, it is all too common for commentators to insist that this must not really be Christianity at all, thereby sacrificing the reality to the preconception.) Yet he is simply drawing upon those same Hebraic and Christian layers, overwritten in modernity but by no means rendered ineffectual, that Traherne had exposed some two centuries earlier. And as Traherne had been to Hobbes, so was Muir to Gifford Pinchot, for whom nature needed to be mastered and "conserved" as a resource and commodity. Muir, on the contrary, looked not to nature as raw material, but to nature as prelapsarian wilderness meant by the Creator to educate humanity by means of its order and wisdom, and above all through its beauty, which he shows throughout his writings to have an unwaveringly sacred character.

Wilderness for Muir is pure, unspoiled, Edenic—the place where it is still possible to walk with God in the cool of the day, as did Adamic, unfallen humanity. Muir's *Thousand-Mile Walk to the Gulf* is replete with language of this sort. Lounging at the side of the Emory River in the mountains of Tennessee, Muir writes: "Every tree, every flower, every ripple and eddy of this lovely stream seemed solemnly to sense the presence of the great Creator. I lingered in this sanctuary a long time thanking the Lord with all my heart for his goodness in allowing me to enjoy it."[28] Several days later he notes: "All the larger streams of uncultivated countries are mysteriously charming and beautiful. . . . Such a river is the Hiwassee, with its surface broken to a thousand sparkling gems, and its forest walls vine-draped and flowery as Eden."[29] And in the swamps of North Florida, he notes that even the alligators and snakes that menace and repel us are true inhabitants of Eden: "Fierce and cruel they appear to us, but beautiful in the eyes of God. . . . They dwell happily in these flowery wilds, are part of God's family, unfallen, undepraved, and cared for with the same species of tenderness and love as is bestowed on angels in heaven or saints on earth."[30] And even among the dead, Muir experiences this divine presence. Surrounded by ancient cypresses and oaks draped with Spanish moss in Savannah's Bonaventure Cemetery, he marvels: "I gazed awestricken as one new-arrived from another world . . . The rippling of living waters, the song of birds, the joyous confidence of flowers, the calm, undisturbable grandeur of the oaks, mark this place of graves as one of the Lord's most favored abodes of life and light."[31] That is, this neglected, human-forsaken place of tombs reveals itself to the pure soul as constituting, in fact, an intermediary realm between the invisible God and visible creation, comparable to that to be found in the angel and saints—*angelois* or messengers between the seen and unseen—and the mundane order of humanity

References like this abound in Muir's writings. In *The Cruise of the Corwin*, looking across the Bering Sea, as it mirrors the lustrous Siberian sky, upon "innumerable multitudes of eider ducks, the snowy shore, and all the highest mountains cloud-capped," he exclaims that, "God's love is manifest in the landscape as in a face."[32] Elsewhere, he notes that although the Yosemite Valley appears to be filled with "a profound solitude," it is in fact "full of God's thoughts."[33] And while returning to his camp in the Tuolomne Valley during the golden days of mid-August, he notes that "bathed" in the glorious, every-changing beauty of mountains and clouds and storms, one is present "at an endless Godful play. . . . Creation just beginning, the morning stars 'still singing together and all the children of God shouting for joy.'"[34] Every reader of Muir is familiar with these kinds of expressions and exclamations: His ecstatic descriptions of nature are everywhere punctuated by them. But they are commonly dismissed in the secondary literature as somehow being merely asides, or literary embellishments. I would like, on the contrary, to suggest that they are rather, for Muir himself, the main point, the periodic summations of a "natural theology" that is very different from that of scholasticism in the medieval West, but rather the re-articulation of a mystical Christian spirituality that his writings are themselves above all meant to convey.

Although Traherne's poetic approach to nature had little influence prior to the twentieth century, since the manuscripts in which these views are most extensively presented remained undiscovered until then, Muir's view of nature has in contrast been extraordinarily influential. His place is truly without challenge, as Chris Highland puts it, as "the patriarch of the American environmental movement."[35] More specifically, his ideas and sensibilities have come to constitute canonically what one of Muir's most perceptive critics, Michael Cohen, has termed the "religion," the "faith or gospel," and citing an oxymoron employed by Joseph Sax, the "secular religion" of the Sierra Club and of what was to become the environmental movement as a whole.[36] So, too, if there is in fact such a spiritual and theological motive that serves as a basis for Muir's work, would this not also serve to call into question the grounding of modern environmentalism almost exclusively on a foundation of natural science and philosophical materialism?

V

The voices of both Traherne and Muir may seem to us extraordinary and anomalous and exotic, standing outside the mainstream of Western thought, deviating not just from what we usually take to be the Christian tradition, but also from the onto-theology that for Heidegger stood as the deepest current of Western thought. But once again, cultural currents turn out to be unexpectedly polymorphous and multi-faceted. The Puritan and Calvinist traditions, upon which both figures draw, are often characterized as dour and world-negating and life-denying, and certainly there are such tendencies within this field of discourse that merit these designations. But there are other strands as well. The Belgic Confession of 1619, for example, with which both Traherne and Muir would have been familiar, maintained that we know God "first, by the creation, preservation, and government of the universe; which is before our eyes as a most elegant book, wherein all creatures, great and small, are as so many characters leading us to contemplate the invisible things of God . . . ."[37] That is, in this most Protestant of documents, we find an epistemology where divinity is not some hypothetical abstraction willfully affirmed by faith alone, a posited entity on the top rung of the onto-theological ladder, but rather we are enjoined here to seek the plain manifestation of the invisible God within our lived experience of the visible world. Moreover, Traherne was strongly influenced by the mystical orientation of the Cambridge Platonists, while Muir was an avid reader of the Protestant natural theology of his day, some of which took a very lively interest in the natural environment.

But I want to argue that there are prominent elements in both Traherne and Muir that cannot be entirely explained by these influences alone, and that evoke even earlier layers of which neither would have likely been aware. To see this, it will be helpful to note some of these common elements explicitly.

VI

First, both operate within not one, but two sets of binary distinctions. One set distinguishes between the visible and the invisible, the physical and the spiritual, or what the Scholastics would have called the natural and supernatural. The other set distinguishes the good and bad, the holy and the unholy, the divine and the worldly, the pure and impure. But it is of the greatest importance to see that contrary to Nietzsche's critiques of both Platonism and Christianity, these dualities do not coincide with one another. Nature not is seen as bad, impure, unholy, but quite the opposite. It is nature itself that is understood to be the pure, the good, and the holy term, and to provide an alternative to the fallenness of the human world, the world of cities and soot, and of the mercantile-industrial society that was at that time being established.

In his autobiographical writings, Muir recalls an incident in Scotland where, as a young boy, he was lowered in a bucket to the bottom of a very deep, and very narrow, well, working to help excavate it. Hitting a stratum that released toxic fumes of carbonic acid and becoming dizzy, Muir looked up into the light and heard his father calling out to him to get into the bucket, so he could pull him out. This becomes for Muir a powerful metaphor, in which the space beneath the well-opening symbolizes the toxic world of cities—toxic not just medically, but above all spiritually; the bucket represents his own writings, as he attempts to pull the reader out into the light and fresh air; and most important, the latter (the world of light and fresh air) is the real physical world, the natural environment itself, but detoxified of the worldly layer, and thus spiritualized, sanctified, divinized. One can find the same sentiments in Traherne's autobiographical reflections, where he emphasizes the importance of his "coming into the countrie"—a phasing of Traherne that was made current in the twentieth century by John McPhee's book of the same title about Alaska—and leaving behind the world of cities and corrupted, fallen humanity.[38]

The second feature shared by both Traherne and Muir is correlative to the first. Both emphasize the need for purification of the soul if one is to truly "come into the country." Both see the need to free oneself from the vice that is typified by urban life, notably but not exclusively leaving behind attachments to wealth and possessions and worldly concerns as a whole—a process that Wendell Berry would someday describe as an essential phase of a meaningful camping trip.[39] Traherne characterizes both of these prerequisites concisely in his "Third Century": "Purity is also a Deeper Thing than is commonly apprehended: for we must disrobe our selvs of all fals Colors, and unclothe our Souls of evil Habits; all our Thoughts must be Infant-like and Clear: the Powers of our Soul free from the Leven of this World, and disentangled from mens conceits and customs."[40]

Third, both writers depict—as following from this purification of the soul—a new power, a new faculty as it were, a new possibility of seeing the physical world as infused with divine energies and numinous beauties. Both see nature, for the purified soul, as itself a realm in which the first set of dualities (visible and invisible, physical and spiritual) are fused, with the physical and natural everywhere shining forth with the spiritual and invisible—itself the ancient, Platonic formula for beauty. And indeed, both see the inability to grasp this Edenic character of nature as the prime symptom of fallenness itself, of the bad, worldly, and impure pole of the second set of dualities. Moreover, what results is a mystical "seeing," and not just an inferential process. The divine is seen within nature, rather than discursively inferred from it.

Fourth, and finally, both relate a consequent involvement with this "Godful" nature, these Edenic landscapes, that they variously describe in terms of joyful immersion and glorious enjoyment. They are not detached spectators, but eager participants, grateful recipients. For Traherne, this means an emphasis upon gratitude, and a reflection upon the ultimate meaning of gratitude, while for Muir it means a total, mystical immersion as he swings atop tall pines in the midst of a cyclone.

But these four characteristics are found a millennium and a half earlier, stated clearly and precisely in the writings of the Desert Fathers and Desert Mothers who not only inaugurated Christian monasticism, but articulated the lexicon of traditional Christian spirituality. For once Christianity had become institutionalized in the great cities of the Roman Empire in the fourth century, the almost immediate response for some was to flee this urbanization and perceived corruption by retreating to the desert, first in the Egyptian Thebaid, and later in the wild highlands of Cappadocia. There these earliest monastics practiced a purification of the soul that correlated to the purity of the place, a process they called *katharsis*. This purification, in turn, allowed them to see divine realities everywhere, and this mystical seeing they called *theōria* or contemplation. More specifically, this allowed them to practice what came to be called *theōria physikē*, natural contemplation or the contemplation of nature: seeing the divine *logoi* within all things. Finally, they wrote of *theōsis*: divinization or union with

God, a joyful and participatory process that at the same time becomes the aperture through which nature itself is divinized. Unlike the Christian West, where it came to be seen in terms of a juridical economy of debt and repayment, in the East, the Incarnation was always seen as a joining together of earth and heaven, of creator and creation, an event silently exerting a divine, synergistic draft upon all of nature and humanity alike.

Remarkably, then, Traherne and Muir—and to varying degrees, other writers and naturalists who come after them—uncover this layer of ancient spirituality that aims at the uncovering of Edenic nature within a corrupted world through a path of inner transformation. I want to suggest that it is the same layer of spiritual sensibility, largely overlain in the West, that is preserved much more thematically in the Byzantine tradition, and articulated eloquently by the Elder Zosima in Dostoevsky's *Karamozov*:

> Love all of God's creation, both the whole of it and every grain of sand. Love every leaf, every ray of God's light. Love animals, love plants, love each thing. If you love each thing, you will perceive the mystery of God in things. Once you have perceived it, you will begin tirelessly to perceive more and more of it every day. And you will come at last to love the whole world with an entire, universal love.[41]

VII

These layers of Western thought, discourse, and experience have undoubtedly been over-layered in multifarious ways by onto-theology—a process that in the West begins most conspicuously with what still seems to me (despite recent attempts at rehabilitation) Augustine's philosophical revulsion at the human body and his denial that God can be experienced within the physical world. But at the other end, it is also over-layed by Heidegger's hegemonic view of history, which effectively ensconces the latter overlay. Although marginalized in the West, these ancient layers of Christian spirituality have been not just preserved, but advanced and culturally inter-layed in the Byzantine East, including Russia and many regions of the Balkans. These may, as well, indicate pointers for articulating a spirituality that is far more promising for environmental thought than the nebulous and ultimately fantastic talk of "gods," in their departure and arrival, that we find in the later Heidegger.

At the same time, however, they also suggest a reappraisal of perhaps the most influential single essay in recent environmental thought, "The Historical Roots of Our Ecologic Crisis," by medieval historian Lynn White Jr. First published in 1967 in the journal *Science*, and reprinted in countless collections since then, this article has persuaded two generations of environmentalists that the responsibility for the current global environmental crisis rests primarily with the worldview of Christianity. Here, White proceeds from a reflection upon a large body of research into late medieval technologies, and concludes that the theology underlying these technologies displays an orientation toward nature that is not just anthropocentric, but "arrogant" in its insistence on seeing human beings as somehow transcending nature, and thus as inherently entitled to exploit it.

Yet those who hastily and uncritically appropriate this critique—seizing upon the supposed injunction in Genesis for humanity to "subdue and dominate the earth," as a proof-text—miss the nuanced character of White's article, particularly ignoring his repeated insistence that his critique applies only to Christianity as it develops within Western Europe, and that matters were quite different in the Byzantine lands. In White's words, "In the early Church, and always in the Greek East, nature was conceived primarily as a symbolic system through which God speaks to men . . . This view of nature was essentially artistic rather than scientific."[42] It is a message and mindset, he adds rather glibly, that the West found "either irrelevant or at best a cabinet of curiosities."[43]

VIII

Yet despite his qualification concerning what he calls the West's "aesthetically magnificent" sister culture of "Byzantium and Russia," White reads Western cultural history as uniform and univocal. This allows White, and later environmentalists, to marvel at the stories of St. Francis consorting with the animals, and to see them as a wondrous anomaly. Whereas in the Orthodox East, no hagiography is complete until some story or other is added about how the saint spoke to the animals and, we may add—inviting comparison with both Muir and his alligators, and to Nietzsche's Zarathustra and his snake—especially to serpents. (The contemporary Elder Paisios, who died only recently, was widely known for the abundance of friendly snakes inhabiting his hermitage on Mt. Athos, even if some skeptics believe he launched this tale himself, primarily to deter the merely curious.) However, the point here is not to valorize a univocal (heroic) Byzantine tradition over an equally univocal (culpable) Latin tradition, but to suggest that there are many more layers embedded within Western culture—and no doubt, all great cultures—than any single, historical-analytic narrative can credibly accommodate, especially when it is a narrative that is as uniform and one-dimensional as those of White and Heidegger—or equally so, those of Vico, and Hegel, and Marx, and Nietzsche, and Derrida as well. All end up telling what becomes a simplistic, and invariably hortatory, tale.

Since the time of the ancient Greeks, Western culture has been multiply porous to other influences, and any attempt at self-understanding would do well to keep this reality foremost in mind. Future chapters will further explore these other layers in Western thought and experience, some of which this chapter has examined in the work of two natives of the British Isles, through a series of comparisons with Eastern counterparts, often chosen from the Byzantine and Russian traditions, but also taken from a representative of the world of Andalusia, fully European and yet for the better part of a millennium one of the great centers of the Islamic civilization, which came to epitomize for Europe "the Orient." But in the following chapter, an understanding of what it means to inhabit the earth is drawn from the Holy City of Constantinople—hinted at, and then ignored by White—spanning over the Bosphorus Straits both Europe and Asia, and according to its inhabitants of more than a millennium, evocatively spanning as well the *chōrismos* between heaven and earth, upon which Heidegger had remained so insistent, and in a way that radically challenges his views concerning immanence and transcendence.

CHAPTER 4

Sailing to Byzantium

Nature and City in the Greek East

> The beauty of the city is not as heretofore scattered over it in patches, but covers the whole area like a robe woven to the fringe. The city gleams with gold and porphyry . . . Were Constantine to see the city he founded . . . he would find it fair, not with apparent but with real beauty.[1]
>
> —*Themistius, fourth-century Byzantine orator*

I

Constantinople. *Constantinopolis. Nova Roma*: "the *polis* founded by Constantine as the New Rome." First established as the Greek colony of Byzantium, it had been settled by residents of ancient Megara, faraway city on the Isthmus of Corinth, the narrow land-bridge between Attica and the Peloponnese. Spanning both Europe and Asia, Byzantium—Constantinople, modern-day Istanbul—has always served as a bridge between these two great continents of the ancient world, a double-headed eagle looking simultaneously east and west. And this was indeed the principal reason for its selection as the New Rome, the future imperial capital for what by the fourth century had become as much an Asian as a European empire. That, and to be a bridge between heaven and earth, a city to do what Old Rome could no longer accomplish: to embody and set to work the ontological bridge between the visible and the invisible at which both occidental philosophy and oriental religion had, in their own respective ways, and to varying degrees, already arrived.

Naturally, it has always been a land of waters that would separate. Waters of the Bosphorus. Waters of the Sea of Marmara—the Sea of Marble. Waters of the Golden Horn. Three waters, everywhere visible, and often audible, ready to isolate it and perhaps, symbolically, to swamp, and overwhelm, the putative city itself in waters and seas. These are not just nearby, for they always exert a presence. They surround and embrace the city, as if to immerse it, inundate it. The city is built down to the very waters themselves, and it everywhere rises up from them.

Civically, it is—on the contrary—a land of bridges. Its bridging and conjoining character is its primary civic feature. Its unity and coherence as a city are functions of human *technē*: both the human art that joins the terrestrial element to the circuit of the city, and the human art that joins the terrestrial with the celestial, the visible with the invisible, the secular with the sacred.

"Therefore I have sailed the seas and come," sings the poet, "to the holy city of Byzantium." Crossing the waters to the holy city, he hopes to find "sages standing in God's holy fire as in the *gold mosaic* of a wall." William Butler Yeats's "Sailing to Byzantium" unfolds with images of golden artifacts and golden artisans, and even golden nature! Goldsmiths and hammered gold. Golden boughs and golden birds. In his prose work *A Vision*, Yeats envisions golden Byzantium as the bridge city, the unifying city, the integral city, reflecting that

> . . . in early Byzantium, [as] maybe never before or since in recorded history, religious, aesthetic and practical life were one, that architect and artificers . . . spoke to the multitude and the few alike. The painter, the mosaic worker, the worker in gold and silver, the illuminator of sacred books, were almost impersonal, almost perhaps without the consciousness of individual design, absorbed in their subject-matter and . . . the vision of a whole people.[2]

He goes on to imagine that had he really sailed to the Byzantium of Justinian the Great, he would have found "in some little wineshop some philosophical worker in mosaic who could answer all my questions, the supernatural descending nearer to him than to Plotinus even." Unity of the religious, the aesthetic, and the practical. Unity of the human, the natural, and the supernatural. The ever-attracting luster of gold, and the mosaic composite of golden fragments that have been drawn together into one.

Finally, in his Preface to the 1893 edition of *The Works of William Blake*, Yeats reflects that

> *In Imagination only* [can] we find a Human Faculty that touches nature at one side, and spirit on the other. Imagination

> may be described as that which is sent bringing spirit to nature, entering into nature, and seemingly losing its spirit, that nature being revealed as symbol may lose the power to delude.[3]

It is thus no longer in Byzantium itself—undone in 1453, as the last act of the genuine decline and fall of the Roman Empire—that we can find the "natural city": the city that would bridge the organic world, the world of the artisan, and the spiritual world, unifying spirit and nature, nature and super-nature. The poet seems to suggest that we can now find the natural city only in the imagination. The bridge—for the modernist poet—is now an interior, and even a psychological function.

Was Constantinople, "the holy city," in fact "the natural city" as well? Let us listen closely to the protocol of witnesses, envoys who sailed to Byzantium in the tenth century. Traveling the earth in search of the religion best suited to unify the Land of *Rus'*, the emissaries of Prince Vladimir of Kiev finally reached Constantinople. Their report back to him is said to have become conclusive for his decision, and it is quoted in every history of Russia:

> Then we went to [Byzantium], and the Greeks led us to the edifices where they worship their God, and we knew not whether we were in heaven or on earth. For on earth there is no such splendor or such beauty, and we are at a loss how to describe it. We only know that God dwells there among men, and their service is fairer than the ceremonies of other nations. For we cannot forget that beauty. Every man, after tasting something sweet, is afterwards unwilling to accept that which is bitter.[4]

As in the poetic narratives of Yeats, in this historic account, too, we find earth and heaven, the visible and the invisible, joined together by means of beauty. But this is not the beauty of nature left in a raw or wild or pristine state. It is that of nature rendered beautiful through *technē*, through human art and artifice: through the art of the architect and poet and iconographer, of ritual and liturgy, and, indeed, through the art of the goldsmith. The beauty that bridges, joins together, and unifies—the beauty that renders possible the seeming paradox of the natural city—this beauty itself comes about not through nature, but through production, through what the Greek language spoken in Byzantium called *poiēsis*.

II

These portrayals of Byzantium find deep resonance in the aesthetic thought of Martin Heidegger, which can help articulate their coherence. In Heidegger's thought, the dark self-closure of earth is thought to be in contrast both to the openness of what he calls "world" as well as to the manifest measure of the heavens. The work of art, then, is understood as "setting-forth" the earth, allowing it to be seen in its earthliness, even as it brings the earthly into the dynamic unity of a world. (And what work of art, we may ask, sets forth the earth more dynamically and dramatically—and sets to work a world more effectively—than the Hagia Sophia, the "Great Church" of Constantinople?) Beauty, in turn, can be seen as the self-revealing or unconcealing of *physis*, nature regarded as what comes forth of its own accord.[5] Moreover, because *physis* or "nature" is not just one region of beings, but rather is everywhere emergent in all that is, to reveal this all-present self-emergence through beauty is at the same time to reveal the unity through which beings as a whole join together and cohere. "Beauty," Heidegger maintains, "is the original unifying One."[6] Nor is this a diaphanous, abstract unity. Because it is "all-presence," beauty is that captivating, *enrapturing* unity that "lets one opposite come to presence in its opposite."[7] (A unity, we may add, that allows the captivating conjoining of visible and invisible, the human and divine, the celestial and terrestrial, sacred and secular, nature and the city.) Finally, Heidegger sees this integral and healing unity disclosed by the arts in their "poetic" character as itself being a revealing of what he calls "the holy." He characterizes the latter, in turn, as the necessary element for humans to encounter the divine, and thus for the authentic poetic task—the task of art itself—to be possible. Perhaps, then, rather than being incidental to its character as the "natural city," the fact that Constantinople was singularly founded *as* a sacred city—as a city that would be the preeminent bridge between earth and heaven, and vice versa—would serve as the very precondition for its singular power to unify nature and humanity as well. The natural city, then, would at the same time be, in the words of Yeats, "the holy city." Byzantium is not only the New Rome: It is also the New Jerusalem—as its residents, in fact, understood it to be.

But we must ask once again, was Byzantium—not just the city of Constantinople, but the inhabited empire itself—in some distinctive and even definitive sense really "the natural city"? What we have considered so far, from the poetic vision of Yeats to the captivated, enraptured report of the Kievan emissaries, is surely suggestive, but it is hardly conclusive. And indeed, there is a body of opinion and scholarship that would argue just the contrary: that rather than being "the natural city," Byzantium represented instead the historical-cultural beginning of the unnatural city.

III

The field of Byzantine Studies is in its early stages, still far from overcoming 1200 years of Western prejudice and provincialism in understanding not just Eastern Christendom, but European history as a whole. The entire period of Late Antiquity—the years from 250 to 800, during which Byzantium took shape and first flourished—has been seen almost exclusively through the prism of the "Dark Ages" undergone in Western Europe after the fall of the First Rome. Eastern Christendom, too, has been viewed from the perspective of the Latin Church that was born out of those dark times. Seen in this way, Byzantine Christianity becomes merely a mystically oriented aberration from the Latin norm, and Byzantium as a whole a curious, rococo remnant, somehow persisting out on the margins. It is hardly surprising, then, that few of those who have thought historically about nature and city have been free from this parochialism. The architectural historian Vincent Scully—who is in other respects without peer for his lifelong study of the relation between nature and the city—does only slightly better than most. Examining his otherwise excellent narrative will sharpen our understanding of Byzantium as the natural city.

In his book *Architecture: The Natural and the Manmade,* Scully presents the fruit of a life's work: a magisterial history of architecture tracing the alienation of the city and nature, from the ancient Egyptians, Minoans, and Mesoamericans to its sad end with the inhumanity of modernist architecture and the frivolity of the postmodern school. Scully sees two tendencies at play in the relation between our buildings and the natural environment around them. One regards the city as part of the landscape and seeks in its architecture to imitate and intensify surrounding nature, to invoke its deities, and indeed to aid and assist them. His favored example is the great pyramid (the "Temple of the Moon") at Teotihuacàn, which Scully sees as mimetically presenting the spirit of the mountain that serves as its background. And in doing so, it evokes—and invokes—the water goddess to bring down the needed water from the earthly heights. (He passes lightly over the fact that all too often, as was certainly the case with the Aztecs, this kind of primal identification with nature has simultaneously entailed human sacrifice to hungry deities as well.)[8] In the second tendency, literally invented by the ancient Greeks, the city stands up against nature, confronts it, raises up human—and, eventually, abstract and geometrical—forms to master and control it. At a certain point, this mastery reaches a point of totalization, in which the world is brought indoors—that is, the interior environment of the buildings becomes the primary element, a world unto itself to replace the natural environment. This latter step, Scully argues, is taken first of all in the Pantheon of Rome, whose dome is conceived as a planetarium, but even more decisively in the great temple, the Hagia Sophia of Constantinople. He sees its circular dome over a square floor plan as the triumph of Pythagorean abstraction—and human control—over the very cosmos itself, and finds that its vastness impresses the visitor with a sense that the building is a world unto itself, subduing and interiorizing the natural world outside.[9]

In part, Scully's analysis simply repeats the time-honored cliché: A world that is old and tired, and that has lost its nerve, now retreats inward—into stoicism and neoplatonism and ultimately into the interior, psychological recesses of Christianity. The Western world begins to waken during the Renaissance and look "outside" to the world of nature, while the Christian East never does, remaining dreamily entranced in mysticism. Yet this doesn't quite work for Scully because the very tendencies he thinks begin in Byzantium in fact proceed more definitively in the West: from the Gothic cathedrals and their interior spaces oriented so decisively heavenward, away from the earth, to the abstract interiorized buildings of the International Style in the twentieth century. But in another sense, Scully occupies solid ground here. His polarity of imitation and identification versus confrontation and abstraction is parallel to Nietzsche's contrast of the Apollonian and the Dionysian, Wilhelm Wörringer's contrast of abstraction and empathy, and indeed Heidegger's notion of a conflict between earth and world.[10] Such a tension, it seems, may be an irreducible element of the human condition. It is the thesis of the present chapter, then, that rather than initiating a decisive fissure between these two tendencies, Byzantium—as the consummation of ancient thought and spirituality—instead presents in an exemplary manner nothing less than a successful resolution of the conflict between them. To see how and why this is the case, however, will require a brief consideration of Byzantine philosophy and theology.

IV

All of the great world religions address themselves to some (typically one) great, ostensibly intractable Problem. For Hinduism, it is the veil of Maya or Illusion—endemic to, and generative of, the very universe itself. For Buddhism, it is Suffering—not just human suffering, but suffering of cosmic proportions—brought about by clinging and grasping. And for Christianity, it is the Fall. But for Byzantine Christianity, and for the Christianity of Late Antiquity generally prior to Charlemagne, the Fall is a disorder of the whole cosmos, of nature as well as humanity. Redemption, then, must in all these traditions have the same cosmic dimensions: a restoration of humanity and nature alike to their prelapsarian condition, transfiguring both nature and humanity, and returning them to their paradisiacal state. In that blessed state, according to the Byzantine vision, human beings would exercise, as they had once before, a cosmic priesthood, apprehending and consecrating the divine presence not only in one another, but in the world as a whole: in every ray of light and each green leaf. The eternal *Logos*, through which the cosmos was created, can once again be apprehended within the inherent *logoi* of all creation because that same Logos entered creation, became material and earthly, precisely to restore this lost unity of heaven and earth. And this allows human beings to once again realize their inherent divinity as images of God. (Byzantine theology calls this process *theōsis*, and it is summarized in the celebrated formula of Athanasios: "God became man in order that man might become God.") Humanity can thus resume the cosmic priesthood for which purpose it was created: to be that being through which the divine image within all creation becomes fully realized, the nodal point through which creation apprehends and consecrates its own inner divinity.

Humanity and nature are retrieved from opposition and confrontation because both are restored to unity with the *Logos* from whom they commonly derive their own being. Because heaven has come down to earth, earth and heaven are now essentially reunited—a theology that underlies all Byzantine art, but which is most characteristically embodied in the art form of the icon. Here, in the icon, the terrestrial is infused with the celestial. The icon, properly understood, is not a representation, but a presentation—not a *Vorstellung*, but a *Darstellung*—of the invisible by means of the visible, a temporal epiphany of the eternal, a visible window upon the invisible. (Latin theology, in contrast, properly begins with the *Libri Carolini*, in which Charlemagne's court theologians—responding to the Second Nicene Council, which had vindicated the icon from the accusations of the iconoclasts—rejected this theophanous character of the icon, insisting instead on the jurisdictional separation of earth from heaven, and substituting the discursivity of allegory and instruction and *analogia entis* for the noetic immediacy of iconic experience.[11])

The background, the very element, of every icon is gold: the inner radiance of the divine energies. That Byzantium is the golden city, that its icons and murals and mosaics radiate with gold, that its ceremonial vessels and reliquaries and garments and gateways are golden, that the pages of its illuminated books shimmer with gold, that its very flag features its double-headed eagle against a golden background simply articulates the Byzantine vision: all this draws toward a restoration of all creation to its divine roots, which can be seen to radiate and well up from deep within the earth itself. It is the golden glow of the pristine dawn of creation shining within the city. But glimpsing the Byzantine flag, which still flies golden over Mount Athos on the Halchidiki Peninsula of Macedonia, let us return to the ancient city itself.

V

Hagia Sophia. The Great Church of the Divine Wisdom. The Divine Wisdom is the eternal *Logos*, seen as shaping the cosmos and holding it together. It is thus also the inner *logos* of each being that, when fully realized, joins it to the whole in a love that must be understood ontologically. St. Maximos the Confessor, Byzantium's greatest philosopher and theologian, states this powerfully: "The unspeakable and prodigious fire hidden in the essence of things, as in the [burning] bush, is the fire of divine love and the dazzling brilliance of his beauty inside every thing."[12] The Great Church of the Divine Wisdom, then, itself serves to bring together all the elements of the cosmos in a transfigured form, making manifest the inner glow of their divine beauty.

Contrary to Scully's Westernized interpretation, the Great Church—like all authentic Byzantine temples—serves not to transport the worshipper to heaven, as the Gothic temple would do, or to replace the natural and earthly with an abstract, Platonized heaven, or even less elicits a psychologized "inner" space. Rather, it serves to join together heaven and earth, to be the ontological bridge between them. The great dome, originally lined with solid gold, still seems to float weightlessly above, as if suspended from heaven or borne by seraphic orders. It is heaven itself, but brought down to earth and joined with it. The Divine Liturgy, for whose sake the church is built, dramatically enacts the joining of heaven and earth: The drama is a progressive interaction and eventual communion of the heavenly (the sacred space and the celebrants in the sanctuary, behind the chancel or iconostasis) and the earthly sphere of the nave, toward whom the icons face, offering the vision of heaven. The rounded apse, deep within the sanctuary, is the cave of Bethlehem, the hollow of earth in which God first assented to become visible. The supporting arches "mark the cardinal directions of space, [and] its piers and pavements the mountains and plains of earth."[13] According to Justinian's contemporary Procopius, the cathedral's "marvelous" and "indescribable beauty" was enhanced by the rich hues of the precious stones in the galleries and arcades, due to which

> one might imagine that one has chanced upon a meadow in full bloom. For one would surely marvel at the purple hue of some, the green of others, at those on which the crimson blooms, at those that flash with white, at those, too, which Nature [*Physis*], like a painter, has varied with the most contrasting colors.[14]

This observation was echoed by a contemporary poet known as Paul the Silentiary, who saw the use of colored marble on the floors and walls as a painting in stone that presented a gathering of twelve kinds of "marble meadows" from the far corners of the earth.[15] And all of this is *oriented*, as is every Byzantine church—that is, it faces the golden glow of the rising sun in the orient, or east.

This could not be further removed from Scully's claim of an abstract, Pythagorean space. Rather, it is much more evidentially mimetic than the Mesoamerican pyramids or the Green Corn Dance of the Taos Pueblo, which Scully valorizes. Yet this *mimēsis* evokes not dark gods, hungry for human blood, but a deity incarnate deep within nature—indeed, a transcendent god become earthly—who nourishes the faithful with his own blood, at the consummation of the Liturgy, under the golden dome of heaven. The Apollonian moment of form and structure, in turn, is not imposed from outside—as a human or mathematized mastery of nature—nor realized as confrontation, but as a restoration of the paradisiacal elements of nature's innermost *logoi*. Aesthetically, it is—in the classical terms revived by Hölderlin and Nietzsche, and in a most unexpected way—a marriage of Dionysos and Apollo.

Much has been made of the de-sacralization of the earth that some claim took place early in the Christian era. This notion forms the basis of Max Weber's influential concept of the disenchantment of nature. Yet this view, as well, sees matters only through Latinized lenses. Long before the rise of Carolingian theology in the West, nature is seen and experienced as iconic, as the visible window upon the invisible. The Byzantine temples and liturgies and holy things set this iconic and noetic relation to nature into play. But it is not just that nature as a whole acquires a new kind of sacred character. It does so locally as well, with regard to specific places. Writing in the Harvard University *Guide to the Postclassical World*, Beátrice Caseau describes a much more complex understanding than the conventional view of "sacred landscapes" in Late Antiquity. Rather than a generalized, pagan sense that nature was somehow sacred, she describes a rich ebb and flow of sacralization, desacralization, and resacralization of specific places in relation to specific deities as peoples and religions migrated and changed. The Christian desacralization was thus a normal part of this process, although accompanied by a new kind of sacralization of place.[16]

Historian Peter Brown has richly documented this process, describing how through holy relics—and, much more importantly, in the East, through the life and death of holy men and women, monastics and saints and holy fools—"paradise itself came to ooze into the world."

> Nature itself was redeemed. . . . The countryside found its voice again . . . in an ancient and spiritual vernacular, of the presence of the saints. Water became holy again. The hoofprint of his donkey could be seen beside a healing spring, which St. Martin had caused to gush forth from the earth. . . . They brought down from heaven to earth a touch of the unshackled, vegetable energy of God's own paradise.[17]

But not only does the ascetic, the holy person of God, sanctify the natural environment through serving as a vehicle of the divine energies: He sanctifies the inhabited places as well. In the Byzantine world of the Christian East, Brown continues, the most important conceptual polarity was not that between city and countryside, but rather between the "world" and the "desert"—and, of course, "desert" in the Orthodox East soon came to refer not literally to the arid expanses of Egypt, Palestine, and Syria, but just as much to the Caves of Kiev and the wild forests of the Russian *taiga* or to the higher elevations of the North Syrian Highlands. The life of the ascetic who can inhabit the wild places of the earth that are usually seen as uninhabitable is angelic in contrast to those more timid and conventional Christians, the *kosmikoi*, who remain "of the world." Such ascetic figures, spiritual athletes themselves, become the most important apertures of all, through which holiness and grace become tangible "in" the world as such, and thus in both the city and the country. Two of the great and exemplary ascetics of early Byzantium were St. Symeon the Stylite, and his precocious successor, Symeon the Younger (521–92), who as a boy set out for the mountains above Antioch. This younger Symeon

> was believed to have played with mountain lions, calling them "kitty." Settled on a high mountaintop, yet still accessible to pilgrims from Antioch and elsewhere, Symeon was believed to have brought back to earth, in his own lifetime, the sweet smell of Paradise, and a hint of Adam's innocent mastery of the animal kingdom.[18]

Those same sorts of wonderful stories that have been told as if they were *sui generis* surrounding St. Francis of Assisi and his empathetic relation to animals and nature—virtually peerless in the Latin West, and extolled by Scully as noble exceptions to the usual relation to nature in Western Christianity—have been regularly observed and recounted innumerable times in the Byzantine East, from the Desert Fathers and Mothers of fourth-century Egypt to the holy hermits of the Russian *taiga* in the nineteenth. Indeed, they are experienced and retold today about not a few of those several thousands of monks still living on Mount Athos, the Holy Mountain. This last, and perhaps greatest, Byzantine holy city—the remote, and nearly forgotten, monastic republic—juts out some 20 miles (32 km) into the Aegean, while preserving intact the religion and culture and sensibilities of Byzantium. Nature on Athos, as visitors invariably report, is indeed holy. Its dozens of cities—monastic communities hanging on cliffsides and clinging to shorelines, merging imperceptibly into the landscape—are strikingly integrated with the natural environment: holy people living close to the land, gathering their sustenance gently and humbly from a landscape that has been sanctified for a millennium. The natural city is resolutely resisting the European Union's insistence that it divest itself of its own "nature" to be "opened up" for mass tourism. For those fortunate enough to have visited and lingered here, it is the strongest evidence of all that ancient Byzantium was, and may remain for us today, in an exemplary way, the profoundly natural city. But for those, too, who have paused perceptively in any one of hundreds of traditional Greek villages—hugging some sea-cliff or nestled in a fertile valley—the natural city of the Byzantines is still alive, if more immediately threatened by modernity and the seductions of European affluence. The novelist Alexandros Papadiamandis, sometimes characterized as Greece's Dostoevsky, has tenderly articulated this ancient village life in a way that constantly shows the subtle bridging of heaven and earth, of secular and sacred, of *physis* and *polis*, in the midst of everyday encounters.[19] And it may even be sometimes glimpsed in the bustle of Greek cities, as a hurried pedestrian suddenly stops to light a candle before an icon at one of Thessaloniki's scores of sidewalk shrines, or as a Vespa rider routinely crosses herself while passing in front of St. Demetrios Cathedral, itself one of the glories of Byzantine architecture.

Whether we regard these glimpses merely nostalgically, as vanishing traces of an archaic past, or rather ponder them more seriously as clues for learning how to build cities that would themselves once again be fully natural—even as we recall Heidegger's insight that learning to build requires learning to once again dwell, as mortals, upon the earth and beneath the heavens, and in the light of the holy—may be decisive for us.[20] For though there are lessons here that can also be learned upon the Acropolis, home to the champion-goddess Athena and her olive trees, while looking west past the port of Piraeus to the oceanic realm of defeated Poseidon—or while looking out contemplatively from the Mount of Olives, across the Kidron Valley, to the still-contested Temple Mount in Jerusalem—Byzantium may nevertheless remain for us in the West, heirs to both Athens and Jerusalem, the exemplary bridge between the secular and the sacred, the temporal and the eternal, between the visible and the invisible: the once and future natural city.

CHAPTER

5

The Resurrection of Nature

Environmental Metaphysics in Sergei Bulgakov's *Philosophy of Economy*

I once heard talk of the old days, how animals and trees and rocks would speak with men . . . In the old days all nature is supposed to have been more alive and full of meaning than in our day.

—*Novalis*

Nature is not what you think,
Not an empty, soulless face,
It has a soul, it has freedom,
It has love, it has a language.

—*Fyodor Tyutchev*

We try to live honorably so that we can discern the inner meaning of existent things, and . . . make our way to the divine Logos in the ontological heart of things.

—*Evagrios of Pontos*

The Depth of Deep Ecology

For nearly three decades, environmental philosophy has been caught upon the horns of a dilemma, bound by the antinomic tension between anthropocentrism and deep ecology. From the beginning in the early seventies, it was clear to many that the "shallow" roots of humanism and environmental anthropocentrism were inadequate. Preserving and maintaining the natural environment for purely anthropocentric reasons seemed not just too limited, but positively wrongheaded. It seemed selfish, mean-spirited—and beyond this, *irreverent and profane*—to see in nature only what it offers humanity, especially since it was the myopic pursuit of self-interest that had forced these reflections upon us in the first place. Alongside the *practical recognition* that we require clean air and water and uncontaminated food to sustain our lives, there had arisen a more tacit, *spiritual awareness* of the need for something much *deeper*: the need for nothing less than a change of heart. This spiritual awareness, then, was fueled by an often unspoken sense of remorse and contrition for damage done to the natural environment—a penitential sensibility whose character is essentially religious—and at the same time a realization that what we had assaulted and jeopardized was not just populations or species or ecosystems, but something deeper, something precious and wonderful: What was endangered was not just places and things, but a principle, power, or *presence* that can only be called numinous or holy. That is, at the root of the sensibility that generated deep ecology—and that still sustains it—there lies an element that is properly religious. And as argued in Chapter 1, it is this element that constitutes the sustaining soil, the very *depth* of deep ecology. It has been the sustaining source of its enduring appeal, but at the same time the reason for much of the hostility that deep ecology has attracted in reaction.

For example, in a 2003 lecture delivered to the Commonwealth Club in San Francisco, novelist and physician Michael Crichton quite unsympathetically assailed the religious character of much contemporary environmentalism:

> Environmentalism seems to be the religion of choice for urban atheists. . . . There's an initial Eden, a paradise, a state of grace and unity with nature, there's a fall from grace into a state of pollution as a result of eating from the tree of knowledge, and as a result of our actions there is a judgment day coming for us all. We are all energy sinners, doomed to die, unless we seek salvation, which is now called sustainability. . . . Eden, the fall of man, the loss of grace, the coming doomsday—these are deeply held mythic structures. [But] the romantic view of the natural world as a blissful Eden is only held by people who have no actual experience of nature. People who live in nature are not romantic about it at all.

Yet one can, he suggests, participate personally in this indigenous perspective. As Crichton urges sardonically:

> Put yourself in nature even for a matter of days, [and] you will quickly be disabused of all your romantic fantasies. Take a trek

> through the jungles of Borneo, and in short order you will have festering sores on your skin, you'll have bugs all over your body, biting in your hair, crawling up your nose and into your ears, you'll have infections and sickness and if you're not with somebody who knows what they're doing, you'll quickly starve to death.

Over against romanticism and religion and mythic vision, Crichton ingenuously commends science and rationality, a cold, hard view of the facts.

> We need to get environmentalism out of the sphere of religion. We need to stop the mythic fantasies, and . . . start doing hard science instead. . . . We [must] manage to get environmentalism out of the clutches of religion, and back to a scientific discipline. . . . Environmentalism needs to be absolutely based in objective and verifiable science, it needs to be rational.[1]

But surely this kind of one-sided, naive reliance on modern natural science begs the question entirely. For it was not romanticism, but the rationality specific to modern science and technology that led to the ongoing, probably intensifying, set of environmental crises that became evident in the last century. How, then, could instrumental rationality by itself lead to its own overcoming? Moreover, as Heidegger has shown, science and technology represent ways of revealing nature that are by no means neutral. To determine nature as object [*Gegenstand*] and resource [*Bestand*] is already normative and prescriptive, especially when this kind of rationally reigns unopposed.

Certainly the humanistic, anthropocentric, rationalistic view that Crichton espouses offers a corrective to what can easily become a misty-eyed, sentimental view of nature. There *is* much in nature that is menacing and hostile. There is much that is repugnant: death, disease, and deformation. Deep ecology too easily glosses this through its reliance on a systems approach—ultimately rooted in the views of Stoicism and Spinozism—legitimizing suffering and disorder with reference to totalities that are supposed to justify them. Yet a one-sided anthropocentrism such as Crichton's quickly degenerates into the Promethean humanism that he shares in common with Bacon, Kant, and Marx. Nature in these quarters is an adversary that must be pursued and cornered, assaulted, and even tortured, to force it to yield its secrets and forfeit its autonomy. What is needed instead, then, is a way to see nature as *both* adversarial and somehow, at the same time, as sacred—a task that will be undertaken later in this chapter.

Finally, like most Promethean humanists, Crichton actually overlooks the most powerful objection to deep ecology. By melding human beings into natural systems, seeing us merely as components of environmental totalities, deep ecology embraces an assumption that is doubly problematic. Questionable, first, because a consistent ecocentrism cannot really regard a human person as no more valuable than a gnat or a worm—or a swarm of gnats and a knot of worms—without either seeming ridiculous, or else killing the claim with a thousand qualifications. But also problematic, because a deep ecologist understanding the interconnections of nature as a whole is no longer just one part of nature among others, but transcends it in the act of understanding. Likewise, the underlying religious sensibility of deep ecology calls for self-sacrifice and overcoming egoism and even a kind of mystical identification with the whole of nature, exhorting us to a higher calling. But of course, a part that sacrifices itself for the whole exhibits the same mode of transcendence as does a part that knows the whole.

Deep ecology has not often sought to make explicit its spiritual roots. To the contrary, the call to live beyond selfishness is often disguised as just good science—as based upon the objective, dispassionate practice of science itself, or else as following from some particular scientific findings. For as Max Weber maintained in *Economy and Society*, the "call" to overcome the passions is itself an ascetic and religious presupposition not only for science, but for the "rationalism" characteristic of modernity as such. Sometimes, however, the religious foundations are noted in the literature of deep ecology—most likely Buddhist or Hindu, if specified at all. Arne Naess goes farther than most by acknowledging the need for deep ecology to be grounded—at the level of what he calls "ultimate premises or ecosophies"—in either a religious tradition (he cites Christianity and Buddhism as examples) or a metaphysical position (citing Spinoza and Whitehead).[2] And in one text, he briefly endorses an openly mystical worldview, prescribing the abandonment of the lower ego-self for the higher "supreme or universal self," a project central to all the great mystical traditions.[3] But more often, the issue is avoided altogether, perhaps because it is subject to assault not only from the libertarian, rationalist right, but just as much from the postmodern left—for whom any reference to concepts such as "depth" and "presence" has until recently been subjected to scorn as indicating a "nostalgia for origins."

This chapter, then, will seek to address both kinds of criticism. It will argue that terms such as "depth" and "presence" are indispensable, as is the impulse behind what is dismissively called a nostalgia for origins. It will also maintain that environmental thought needs to strengthen and articulate its underlying theological orientation rather than purify itself of it. It will even argue that mythic terms such as "paradise" and "fall" are not only inevitable, but richly helpful in understanding the relation between humanity and nature. Thus, I believe that overcoming the false dilemma between anthropocentrism and deep ecology requires movement in this direction. Whether we begin with humanity as transcending nature or as immersed within it—and whether we begin with nature as fallen and in need of rectification, or as paradisiacal and deserving of veneration—either side pursued rigorously and consistently will lead to an underlying theological dimension that we shall term here, after Bulgakov, "sophianic." Moreover, not only does it underlie each respective position, but if understood correctly it would reconcile it with what appears to be its opposite.

I also believe the recent history of environmental philosophy leads in the same direction. Starting out as one branch of "applied ethics"—alongside business ethics, medical ethics, and so on—the need soon became apparent to rethink the very notion of the ethical, leading to a deeper understanding perhaps best exemplified by the "dwelling" of Heidegger's later philosophy. And it is now becoming evident that environmental philosophy must reach beyond the ethical altogether: that we need an environmental aesthetics alongside environmental ethics—in Heidegger's language, that "dwelling" must be "poetic." But of course, for Heidegger, poetic dwelling is household inhabitation that is beneath the heavens and receptive of divinity, awaiting and making space for the advent of the holy. Other phenomenological approaches have arrived at similar conclusions, and virtually all the great nature writers have found something numinous in nature's beauty. The Czech philosopher Erazim Kohák has written not only one of the best studies of Husserl's *Ideas*, but also a great, neglected masterpiece of environmental philosophy, *The Embers and the Stars*, published in 1984. Here, Kohák argues brilliantly that understanding nature as it really presents itself to lived experience requires not just Husserl's phenomenological bracketing of the natural attitude and of intellectual constructs concerning nature, but a *practical epochē* that would bracket the technological artifacts and systems that conceal lived nature from us, even more effectively than the conceptual constructs.[4] Hence, this radical—phenomenological *and* practical—bracketing reinstates a numinous vision of nature, one that is far closer to mythic consciousness than to the scientific consciousness of the ecologist. This should not be altogether surprising, for the thinker who pursued this path perhaps most persistently—Henry David Thoreau—noted in his journals that the only writing to give "any adequate account of that Nature with which [he was] acquainted" was the literature of "mythology"—"as old as mankind."[5] And Kohák drives home his point eloquently. Through this twofold, radical bracketing, nature re-emerges not as dead matter, but as gift and givenness: as creation. "In lived experience," he writes,

> in the radical brackets of the embers and the stars, the presence of God is so utterly basic, the one theme never absent from all the many configurations of life's rhythm. The most basic trait of the world that confronts a dweller in the radical brackets of the forest clearing is that it is God's world, not "man's," and that here God is never far . . . Nature's gift to humans is . . . the power and grace of God which confronts the dweller therein. That is the most basic given, the gift of the dusk.[6]

Kohák's book anticipates the recursion of the religious that has been intensively debated in French philosophical circles, and is centered upon the debate between Jean-Luc Marion and Jacques Derrida concerning the possibility of a phenomenological approach to "the given." I refer to this as a "recursion" of the religious not simply to indicate a recurring, but to suggest the mathematical sense of recursion as well: the possibility of a set being defined in terms of another set, which is likewise defined serially in terms of a third set, and so on potentially to infinity. As the twentieth century saw repeatedly, religious predisposition recurs in just this way, reappearing recursively in new forms that reiterate it. George Steiner, in *Nostalgia for the Absolute*, shows convincingly how not just irrationalist fads such as astrology and occultism, but Marxism, Freudian psychoanalysis, and Levi-Straussian anthropology, all display systematically the recursive characteristics of religion, albeit in less successful packages than traditional modes. But this recursion of the religious is even more striking in the nature writers of the nineteenth and twentieth centuries. Just when the Enlightenment seemed to have purged nature of theology and reduced it to a mechanical system, the religious recurs: first in European romanticism and American transcendentalism, and then increasingly in a variety of writers who carry through something akin to the radical bracketing Kohák prescribes, and arrive at a similarly radical finding: a discovery of holiness, divine presence—of what traditional theology called "the glory of God"—in the beauty of nature. From Muir and Burroughs to Annie Dillard and Barry Lopez, we consistently find a literary passage from environmental aesthetics to environmental theology. It can even be found in Nietzsche's discovery in the landscapes of the Engadine Alps—hiking six thousand feet above man and time—of "eternity, deep, deep eternity." There are, I will argue, important philosophical and theological grounds for this recursion of the religious—or if one prefers, of the mystical. To examine them, we will turn to a thinker whose work has powerful implications for environmental thought overall, and for resolving the dilemma of anthropocentrism and deep ecology, as well as several others. But first, it will be important to briefly consider the rather surprising terrain of Russian thought, which the careful reader will find is far less familiar than it might be assumed, and which will offer background not only for this chapter, but for the later discussions of Dostoevsky and Florensky in Chapters 8 and 9. As should become apparent, traditional Russian culture—in contrast to the sterile dialectical materialism of the Soviet Era—offers us perspectives that exhibit simultaneously both a profound identity with Western thought and a radically different set of assumptions and sensibilities (and indeed, compatibilities, such as between radical mysticism and tough common sense) that can be especially powerful in helping us look beyond dichotomies in which we have become stuck, such as that between anthropocentrism and deep ecology.

The Nature of Russian Philosophy

Throughout most of the twentieth century, both the rich cultural heritage of Russia and its profound effects upon the West have been eclipsed—inside as well as outside Russia—by the shadow of the totalitarian government to which it had itself given birth. Only recently is it beginning to be recognized that the influence of Russian culture upon the culminating period of modernism, and its transition into post-modernity, has been so decisive that historian Steven Marks has plausibly entitled a recent study *How Russia Shaped the Modern World*.[7] For it was the powerful current of anarchist theory running from Bakunin to Kropotkin that gave to the world in their first forms both ecological utopism and modern terrorism, even as the writings of Tolstoy first systematically articulated the theory of nonviolent resistance. The concept of "nihilism"—along with the nihilist sensibility—was born in Russia as well. Tolstoy and Dostoevsky influenced modern literature so profoundly that George Steiner has claimed only Attic tragedy and Shakespeare have had a comparable influence. Russian painters such as Kandinsky and Malevich, "purists of beauty," gave the world its first abstract paintings, going beyond three dimensional art. In the performing arts, figures such as Chekhov, Nijinsky, Stravinsky, and Stanislavski were decisive in the rise of modern drama, ballet, music, and acting theory. Occultist and New Age sensibilities were first synthesized by Madame Blavatsky living in London. And in France, the intellectual history of science, and the Hegel most influential in the twentieth century, were shaped by the Russian emigrants Koyre and Kojeve. And of course, the totalitarianism pioneered by Lenin, and perfected by Stalin, was emulated in various forms by admirers as diverse as Kemal Ataturk in Turkey, Hitler and Mussolini, Mao and Ho Chi Minh, as well as Nasser of Egypt and the ideologies of the Middle Eastern Baath Party.

But just as impressive, although thus far less influential, have been the achievements of Russian philosophy, an intellectual milieu that largely remains *terra incognita* not only in the West, but until recently in Russia itself due to its suppression by the Bolsheviks. The period of philosophical activity beginning with Soloviov in the late nineteenth century and culminating with Florensky, Berdyaev, Shestov, and Bulgakov in the twentieth has few rivals in the history of philosophy for its comprehensive scope and its intellectual power. It is a philosophical tradition that has great appeal and great promise for the practice of philosophy today, drawing as it does on the deep springs of German Idealism and Neoplatonism, while avoiding the pitfalls of dogmatic metaphysics through a phenomenological emphasis on lived experience, an orientation toward socio-political *praxs*, and above all a concrete rootedness in everyday spirituality based on the mystical tradition of the Christian East. And perhaps one of the best ways to gauge the brilliance of this philosophical milieu is through seeing its dazzling implications for the incipient discipline of environmental philosophy. There are numerous reasons for this, but four are especially salient.

First, there is a deep tradition of nature mysticism reaching back to Russia's pagan past, but just as strongly rooted in Byzantine and Russian Orthodox spirituality. For example, "Hagia Sophia"—the holy wisdom of God, the divine wisdom coursing throughout the cosmos—lends its name not only to the Great Church of Constantinople, but in its Cyrillic form names great cathedrals throughout Russia and Ukraine, renowned icons of the Russian Church, as well as liturgies and feast days. It has often been seen as the idea most characteristic of Russian thought and spirituality, and it allows for an understanding of the divine immanence in nature that is neither pantheistic, nor even panentheistic, for it is solidly based on the Byzantine distinction between the divine essence or *ousia*—eternally hidden and unknowable, even to the celestial orders—and the divine energies or *energeia*, immanent throughout creation and noetically present to the holy and to the chosen. Environmental philosophy, as I have maintained already, has great need of an understanding of nature in its holiness—an experience that is attested in innumerable nature writers, but which has been left without conceptual articulation by philosophers and theologians alike.

Second, and closely related to this, the role of beauty is always central in Russian culture—in its philosophy, its literature, and its theology. It was, after all, Dostoevsky who famously prophesied that it is beauty that will save the world. But this love of the beautiful, this *philokalia*, reaches far back to the early medieval envoys of Prince Vladimir of Kiev, who traveled the world in search of a religion suitable for the pagan *Rus'*, and who were overwhelmed by what they encountered at the Church of the Hagia Sophia in Constantinople:

> Then we went to Byzantium, and the Greeks led us to the places where they worship their God, and we knew not whether we were in heaven or on earth. For on earth there is no such splendor or such beauty, and we are at a loss how to describe it. We only know that God dwells there among men, and their service is fairer than the ceremonies of other nations. For we cannot forget that beauty[8]

As Plato and all Platonism has maintained, it is in beauty that the eternal breaks through into the temporal, that the forms are actually perceptible, that invisible becomes visible. Or as the Byzantine-Russian tradition would state it, it is in beauty that the divine wisdom, the holy wisdom through which all the cosmos is created and upheld and renewed, is most characteristically experienced. And I suggest that something of this order is an indispensable element for a viable environmental aesthetic.

Third, the Enlightenment came late to Russia, and when it did, the triumphalism of science and discursive rationality met not with a relatively diminished and debilitated spirituality as in the West, but with a more intact and vigorous spiritual tradition that could hold its own against the strong current of instrumental rationality. This has, I believe, given Russian philosophy an ability to maintain a freer relation to the scientific view of nature, both subjecting it to a more radical—and less reactionary—criticism than in the West, while at the same time appropriating it when suitable in a guileless and unassuming manner that has seemed untenable to Western critics of science from Blake to Heidegger. And surely there is no task for environmental philosophy more urgent than a free, and creative, encounter with modern nature science, whose view of nature has so far been uncritically appropriated by most environmental thinkers and theorists.

Finally, as I will discuss in more detail later, Russian philosophy draws heavily upon a cosmic eschatology that sees nature as awaiting its own resurrection through the redemption of humanity—that is, through the redemption of the very agent through whom nature is fallen and the divine wisdom inherent within it becomes obscured. This allows environmental thought and practice as a whole to be contextualized—as seems increasingly demanded by global realities such as climate change—within a cosmic struggle for the triumph of the motley and the various and the beautiful, of the hale and holy and the healing over corruption and disfiguration, of life over death.

We may return, then, to this vexing conceptual dilemma that has impeded the development of environmental philosophy: that between anthropocentrism and deep ecology, between the view that the meaning and value of nature is rendered through its relation to the human sphere, and the view that human beings must be understood—and more important, must understand themselves—primarily as members of a larger biotic community. It is against this background that we can most usefully approach the philosophical work of Sergei Bulgakov, whose early work—upon which this chapter will focus—while lacking much of the later nuance and sophistication of his later work, is perhaps most commensurate with Western philosophical modalities. The work of his friend and colleague, Pavel Florensky, who has been briefly mentioned previously, has equally important implications for environmental thought, and his thought will be examined later (in Chapter 8).

Household Life

Until recently, the work of Sergei Bulgakov was suppressed in Russia and largely untranslated in the West. But as this neglect gets redressed, it is becoming evident that Bulgakov—along with figures such as Shestov, Florensky, and Berdyaev—was one of the greatest Russian philosophers of the twentieth century, and it seems even clearer that in the wide-ranging originality of his thought as a twentieth-century theologian, he is perhaps without peer in the Orthodox East, and perhaps the West as well. He started out, however, as an economist holding a prestigious chair in agricultural economics—trying to adapt Marx to the largely agrarian economy of Russia—and as a politician who served in the second Duma, or Russian Parliament. Along with many Russian intelligentsia during the first decade of the twentieth century, Bulgakov—after publishing in 1900 a major work called *Capitalism and Agriculture*—became disillusioned by the inapplicability of Marxist economics to Russian reality, as well as disenchanted with the irresponsibility of the left in the 1905 Revolution and subsequently, in the Second Duma. Together with Berdyaev, he edited in 1909 an influential collection of articles—much excoriated by Lenin, living in Switzerland—called *Vekhi* or *Landmarks: A Collection of Essays on the Russian Intelligentsia*, which sought to articulate a more indigenous, progressive vision for Russia than what the intelligentsia now largely regarded as the failed revolutionary politics. Intellectually, Bulgakov moved first from economics into philosophy, and then from philosophy into theology, eventually being ordained a priest in the Russian Orthodox Church in 1918. He was exiled from Russia by the Bolsheviks in the early twenties, and continued his academic life in Paris until 1944 when, according to some present, he died as a saint. Published in 1912, *The Philosophy of Economy: The World as Household* represents Bulgakov's major work as a philosopher, finalizes his break with Marxist political economy, and establishes the philosophical grounds for his later work as a theologian.

The book opens by acknowledging the inescapability of Marx's central insight: the "economism" or "economic materialism," which despite its reductionistic assumptions, has become "the reigning worldview" of our time.[9] Bulgakov grants immediately, then, the *truth* of "economism," the "profound content which shimmers through" economic materialism, its "indestructibility"—despite its one-sidedness and abstractness. Writing in 1912, he foresaw what became widely evident only later: that a central element of our current historical experience is the sense that "*life is, above all, an economic process.*" "Our time," he continues, "understands, feels, experiences *the world as a household*—i.e., in Greek, the world in some undeniable sense now presents itself to us as *ecos*."[10] In contrast to the later Heidegger, who recoils at the global, technological disclosure of nature as inventory (*Bestand*), Bulgakov grasps its inevitability and prescribes not Heidegger's step back, the *Schritt zurück*, but seeks instead to tunnel into this contemporary experience—to show internally its one-sidedness and dogmatism, stating that economism should be "overcome from within," even as he acknowledges its *partial* legitimacy. Indeed, he even maintains that "not to experience" this "peculiar enchantment of economic materialism" at all, "not to feel its hypnosis (even if one does not abandon oneself completely), means to have some defect of historical consciousness, to be internally alien to contemporary reality," artificially aloof.[11] And in contrast to the environmental purist, seeking enclaves of pristine nature unsullied by humanity, Bulgakov notes from the beginning that *of course* all nature has become for us today an economic product. But he goes on immediately to show that this economic status entails far more *ontologically* than it might appear. And thus begins a reversal from within.

A reader coming to Bulgakov with a background in phenomenology will be struck by how many of the central themes of Husserl, Heidegger, and Merleau-Ponty are anticipated, sometimes by half a century. For example, just as Heidegger was later to begin *Being and Time* with a unitary thought (*Dasein* as Being-in-the-World) that dramatically subverted the bifurcation of subject and object, so Bulgakov inaugurates *The Philosophy of Economy* with an understanding of economic activity that not only destabilizes the rigidity of economic materialism, but also undermines the later antinomy between anthropocentrism and ecocentrism. The philosophy of economy, then, will begin from a sense of "philosophical wonder"—a sense of strangeness like that undergone by a visitor from a faraway land, for whom nothing is customary or conventional—at the defining problem of economy: "humanity in nature—and nature in humanity."[12] This philosophical wonder will be double, yet respond to something single and unitary, to what in Russian is called *koziaistvo*, a word denoting economy and economic activity, but even more basically household life, home management, house maintenance—thus preserving in its Russian root, the same meaning as the Greek *oikos*, the root for both economy and ecology.

The world is a *household*, and this insight opens up the philosophical perplexity that eluded Marx, preventing him from understanding the true character of economy. Just as Heidegger, in his essay "Building, Dwelling, Thinking," much later sought to show phenomenologically that the genuine problem of housing in postwar Germany was not primarily logistical, so Bulgakov seeks to elucidate how the economic problems of late Czarist Russia—equally concrete issues, such as land reform and peasant proprietorship—could be understood adequately only from within, through a study of the lived dimensions of economic reality, and ultimately through an ontology of economic activity that stayed close to its phenomenological moorings in the lived experience of the *khoziain* (the proprietor or subject of economic activity—both as individual, and more importantly as collective) interacting with nature (the object of economic activity). This interaction is what constitutes *khoziaistvo*—economic activity as household life.

Thus, Bulgakov rejects Marx's scientific objectivism as dependent on Kant's understanding of nature as object, and as leading to the view of labor as the effect of one object upon another. In contrast (and this has lent his work tremendous urgency in the reforms of post-Soviet Russia), Bulgakov insists on understanding *koziaistvo* (economic activity, household life, or simply housekeeping) from *within* the lived experience of the *khoziain*, the householder, the head of the household, the host (to the visitor), the proprietor in the sense of the proprietor of an inn, or as he will even call him later in the book, the world steward and even the demiurge. The central perplexity of economy, then, is not labor, seen as the external effect of one object upon another, but something much more remarkable and perplexing and wonderful, worthy of true philosophical surprise (*thaumazō*): It is the world as household. Nature as a Kantian "object" is an abstraction. So too, correspondingly, the genuine "subject" is not what Bulgakov calls "the Kantian epistemological subject"—"the mind of a scientist preparing an experiment"—detached, and disembodied.[13]

Like Heidegger, Bulgakov maintains that nature is more primordially encountered within practical activity (as *Zuhandenheit*) than in theoretical consciousness. But as Heidegger often seems to overlook, practical activity presupposes our being part of nature: "In order for economy to be possible, the subject—the world proprietor, or demiurge—must be part of the natural world, immanent in empirical reality."[14] Thus Marx, and the whole tradition of materialism, and the biocentrists are all correct here: Humanity is a part of nature, *humanity in nature*, a housekeeper and an inhabitant. Nature is not, however, a Kantian object, but a *household* ripe with human meaning; it is always already in its very givenness *humanized*, and science is itself part of that humanization from within, not a disengaged mirroring from outside. Thus, the anthropocentrist and humanist are also right—and Heidegger is right—in maintaining that the dweller and householder, who alone can make the world a household, cannot be just another part of a greater totality of objects, but must somehow lend meaning to the whole.

The primary datum—both phenomenologically and ontologically—is economic activity as household life, something utterly concrete, something integral and cohesive that must yet be understood both as humanity in nature *and* as nature in humanity. What is needed, then, is an ontology of the economic process, understood from within the lived experience of the householder.[15] "Every economic act," Bulgakov writes, "realizes a certain fusion of subject and object, the penetration of the subject into the object, the *subjectification of the object*—or the subject's exit from itself into the world of things, into the object, that is, an *objectification of the subject*."[16] Humanity in nature, and nature in humanity. Objectification of the subject and subjectification of the object. Biocentrism and anthropocentrism. Two aspects abstracted from something concrete and integral, two faces of a single reality. Bulgakov makes explicit here his philosophical debt to Schelling's philosophy of identity, for which "nature must be visible spirit, and spirit must be invisible nature."[17] The starting-point for Bulgakov, however, is not the critical idealism of Kant, but the economism that subtly underwrites modern science and advances its domination of environmental thought.

For the ground of *scientific* materialism, Bulgakov argues, is *economic* materialism. And this economic underpinning holds for biocentrism as well, to the extent that it rests not on a vision of mystical identity—toward which deep ecology tacitly inclines—but on the results of positive science. The seemingly passive scientific subject is an expression of the active economic subject. It is not its mirroring by an epistemological subject that constitutes the reality of nature: "The entire practice of mutual interaction of I and non-I establishes the reality of the external world and fills the empty and cold realm of the non-I with strength, warmth, bodies, turning the *mirage* of the non-I into *nature* and, at the same time, placing the I *within* nature, organically fusing them into a single whole."[18] The knower is neither a mirror nor a phantom, but the subject of a mutual interpenetration of humanity and nature. But who *is* this subject of economic activity?

The interaction and interpenetration of humanity and nature—nature as household and ourselves as proprietors—presupposes a unitary field that must be always already in place and in play, prior to any individual engagement. Economic activity, the world as household, has a hereditary and historical, as well as a social and collective, character. "*Economy* as a whole," argues Bulgakov, "is not only logically but empirically prior to separate economic acts."[19] Who, then, is the subject of economy as a whole? "The single true transcendental subject of economic activity, the personification of *pure economy*, is not any given individual but *humanity as a whole*."[20] Ontologically, Bulgakov understands this notion of humanity as a whole, of a universal subject, in a robustly realist manner, not as an abstraction or universal, nor as a methodological device, as the transcendental subject was for Kant. For economic acts to cohere into a system—and therefore for cognitive acts to cohere as science, and for human actions to come together as history—human knowers and agents must not be impermeable to one another, but function as nodal points for humanity as a whole.[21]

"There is one subject," Bulgakov maintains, "and not many: the transcendental subject of knowledge, of economy, of history is clearly one and the same; it founds and objectivizes all of these processes, transforming the subjective into the transsubjective, synthesizing the fragmented actions and events that make up economy, knowledge, and history into a living whole."[22] (Bulgakov regards Kant's individuation of the transcendental subject as a "mystical misstep," an error by means of which Kant "reflects the fundamental sin of Protestantism" by positing the will and consciousness of the individual in opposition to the "supraindividual unity" of humanity.[23]) This does not, of course, mean that individual persons are somehow unreal, but simply that when they engage in what ontologically establishes their own humanity, they do so by taking part in something shared and common: "*Only one truly knows, but many engage in the process of cognition.*"[24] This collective, universal humanity is none other than what has been called since antiquity the world-soul, and whose lineage Bulgakov traces from Plato, Plotinus, and the ancient Stoics through Sts. Dionysios, Gregory of Nyssa, and Maximos in the Greek East—and Scotus Erigena in the Latin West—to Böhme, von Baader, Schelling, and Soloviov in modern times. Precisely in its character of *khoziain*, of world proprietor, humanity is at the same time world-soul, the very *eye* of the world soul, the world soul become hypostatic—nature acting upon itself, recognizing and realizing itself through that agency. And the aim of this momentous movement, the goal of that self-realization, is a permeation of nature—in all its seemingly lifeless recesses—with humanity, an infusion that is just as complete as the manner in which nature permeates humanity in all that seems most purely ideal. That reciprocal interpenetration is not just agriculture and industry. It is precisely what knowledge accomplishes. And even more importantly, it is preeminently what is undertaken by art—the perfection of economic activity understood as household life, the completed self-realization of nature in the element of beauty.

Sophia

But the highest expression of economic activity is neither factory nor farm. Bulgakov maintains that the epitome and perfection, both "the goal and limit" of economic activity—i.e., of household life—is *art*, through which human creativity transforms and transfigures nature by uncovering and unleashing its beauty. He even argues that "economy [as such] must return to its [Edenic] prototype, must become transformed into art."[25] Thus, those who today argue that nature is a social construction are partially right. Through art, as through science, as through traditional practice and historical experience, nature emerges transformed as a result of human labor. But to see *only* this is to espouse the same kind of one-sidedness as the environmental positions discussed earlier: materialism, anthropocentrism, biocentrism, and so on.

The intellectual historian Marjorie Nicholson has shown how our appreciation of the beauty of wild landscapes is a rather recent acquisition. For example, she documents that until the last few centuries, the Alps were regarded as so repulsive in their bleak, chaotic, barren disorder that coaches traversed them with curtains drawn, to protect travelers from such a repugnant spectacle.[26] The wild nature we work to preserve in the harsh mountain high-country, the scoured canyons of the American Southwest, and the flooded marshlands and estuaries of the American Southeast is beautiful, but it is a beauty that is, in an important sense, a creative product. A product of what? Two centuries of romanticism in poetry, music, and painting. The philosophical concept of the sublime. A century and a half of nature-writing, from Chateaubriand and Bartram and Thoreau, to Dillard and Lopez. Dazzling historical accounts by amazed explorers and naturalists from Powell and Muir to the present day. Photographers such as Adams and Weston and Porter, plus a half-century of Sierra Club calendars and coffee-table books. The point is that the wild nature we find beautiful is a humanized nature, the result of economic activity, a product of human art and creativity. Nature in humanity—the object subjectivized. Or as V. S. Naipaul put this, the work of the imagination "hallows it subject," such that "landscapes do not start to be real until they have been interpreted by an artist."[27] But does that mean that the result is a fiction or fabrication—somehow unreal? To the contrary, it is far more real, Bulgakov argues, than the nature revealed by scientific theories from Copernicus to the present day, theories that are just as much creative products of human economy, and which are hardly invalidated by that status.

For science and art and the simplest economic activities of toolmaking and agriculture all proceed from the same source: the discovery—within nature as a lifeless, mechanical collection of blind forces, an inanimate realm of brute necessity, within *natura naturata*—of vital interrelationships, of an unbounded field of reciprocally connected and mutually penetrating forces, a "logos of things," which we ourselves engage as both participants and revealers. Evoking the Stoic thought of the *logos spermatikos*, Bulgakov later writes, "The truth is that nature is not empty, but full. It is full of *logoi*, ontic seeds, which pre-contain the *all* of cosmic being."[28] In short, we find everywhere incipient life, organic interactions and the palate of the beautiful, but frozen in immobility and awaiting liberation through our knowledge and creativity. That is, we find *natura naturans*, discovering once again the world soul—this time not in humanity as transcendental subject, but in nature as transcendental object. The conditions for the possibility of economic activity of any kind—the ontological ground for the world as household—lie in this self-recognition of humanity in nature, and the simultaneous self-realization of nature in humanity. In economic activity, the life of the earth household, we thus find a twofold manifestation of the world soul: as realizing the potential for collective activity of the transcendental subject, and as realizing the potential for universal life and freedom and transfiguration in the transcendental object. Discovery and creation are not two different practices here, but aspects of a single process of *self-realization*—to use a term favored by Arne Naess, even as he struggles to justify it philosophically. What happens, however, if economic production is merely willful and arbitrary? What if putative creativity results not in the animation and transfiguration of nature, but in its distortion and debasement and degradation. In Blakean terms, what if a perverse economy results not in "mountains green" and "pleasant pastures seen," but in "clouded hills" and "dark satanic mills"? And why the earlier claim, still not justified, that nature as it is revealed by art is truer, more real, than the nature revealed by science?

To answer these questions, we must consider how Bulgakov develops two themes that are central to the Byzantine philosophical and theological traditions, and in particular to their articulation in Russian Orthodoxy—concepts somewhat dissonant with Western thought in the last millennium, but deeply resonant with the shared experience and traditional wisdom of non-western humanity. *First*, in the Christian East, the concept of the Fall is understood to be not exclusively a fall of humanity, but of the cosmos as a whole, due to the waywardness of humanity. It is not just we ourselves that are fallen, but nature too, because of us. Accordingly, it is not only humanity that stands in need of redemption, but all of nature—whose inherent goodness and beauty has been afflicted with discord and corruption. *Second*, underlying the thought of cosmic fall and redemption is the concept of Sophia—the Divine Wisdom elaborated at great length in the Wisdom Books of the Septuagint Bible, and integral to Eastern Christendom since the fourth century building of the great Hagia Sophia, the Church of the Holy Wisdom, in Constantinople, and its namesake successors in Kiev and Novgorod. Sophia is that element—manifested in *natura naturans*—through which nature in whole and part is rooted in the Divine being, and which thus provides the normative measure within nature for what it once was paradisiacally, and for what it once more must be through human creativity. And thus it provides the standard for what Bulgakov calls an Edenic economy, household life that restores and redeems and indeed transfigures cosmic nature.

Speaking of itself—or indeed *herself*, as the passage is gendered in the Septuagint Proverbs—Sophia, the Divine Wisdom, the Eternal Logos interacting and infusing creation, states, "I was set up from everlasting, from the beginning, or ever the earth was. When there were no depths, I was brought forth; when there were no fountains abounding with water. Before the mountains were settled, before the hills was I brought forth . . . When [the Lord] prepared the heavens, I was there; when he set a compass upon the face of the depth . . . when he appointed the foundations of the earth; then I was by him, as one brought up with him; and I was daily his delight, rejoicing always before him; rejoicing in the habitable part of his earth; and my delights were with the sons of men" (Prov. 8:23–31).[29] And in the Book of the Wisdom of Solomon, Sophia is "she who is the artificer of all things," who "pervadeth and penetrateth all things by reason of her pureness. For she is a breath of the power of God, and a clear effluence of the glory of the Almighty . . . an effulgence from everlasting light and an unspotted mirror of the working of God, and an image of his goodness" (Wisd. of Sol. 7:21, 24–26).[30] Surely, as Bulgakov notes, Sophia has certain similarities to "the world of [forms] discovered for philosophy by Plato," and Bulgakov somewhat misleadingly anoints Plato as the "prophet of Sophia."[31] For Bulgakov is by no means a Platonist, and Sophia is just as much what Heidegger calls "the holy" and "divinely beautiful nature" in his Hölderlin essays, not something transcendent, but nearer the surface, bubbling over in the beauty that allows us to apprehend nature's rootedness in the saving and healing power of the holy. It is that to which Dostoevsky referred, in writing that "beauty will save the world"—or in Bulgakov's words, it is "beauty as that unceasing force that strives within every being towards the realization of its own *logos*, its eternal life."[32] It is, argues Bulgakov, "the spirituality of nature . . . the grace of the Holy Spirit that inheres in it," "that ineffable and rationally unfathomable beauty which delights, nourishes, freshens, and fills the soul."[33] Sophia is Hopkins's "dearest freshness deep down things" that can nevertheless "flame out like shining from shook foil," and Sophia is the nature characterized by Thoreau as "Nature, lying all around, with such beauty . . . a personality so vast and universal that we have never seen one of her features."[34]

"Sophia shines in the world," writes Bulgakov, "as the primordial purity and perfection of the universe, in the charm of a child and the enchantment of a fluttering flower, in the beauty of a starry sky or of a flaming sunrise. . . . These sophic rays," he continues, "are what attract us to nature."[35] Sophia is revealed preeminently in the beauty of nature, which Bulgakov calls both "the vestment of divinity" and "the flower of creation"—"that ineffable beauty which delights, nourishes, freshens, and fills the soul . . . [that] *spiritual force* that testifies about itself to the human spirit."[36] And it is also, of course, Sophia that allows us to sort out what is Edenic from what is fallen—and to distinguish the Edenic economy which actualizes the latent life and beauty and goodness of nature, from an economy that Bulgakov does not hesitate to name, along with Blake, diabolical: "If [economic] creation takes matters into its own hands, seeking a model outside the divine Sophia, it shapes a shadowy, satanic world alongside the given, created one."[37] The Edenic economy, in contrast, he defines as "the selfless loving effort of man to apprehend and to perfect nature, to reveal its sophic character." This corresponds to the "Edenic state" in which economic activity began, in which the household was itself Eden, and in which humanity was created to be "the living tool of the divine Sophia," to name the birds and animals, to transform the whole world into the garden of Eden, and in the words of Genesis, "to tend it and keep it"—in which, "originally, economic activity *was* the harmonious interaction of man with nature."[38]

But why revive these old thoughts of paradise and fall, even though Bulgakov states clearly—as does Kohák, who draws upon them as well—that he is drawing upon them not as mundane history, but phenomenologically and ontologically, "according to a certain empirico-mystical geography?"[39] Do they not affront modern science so egregiously, that even those critical of scientism must find them far-fetched? Isn't everything in nature just . . . *natural*? Perhaps the simplest answer is that this ancient view of nature seems not only unavoidable phenomenologically—and it seems to me that Annie Dillard has made this case very strongly in her writings on nature, especially in *Pilgrim at Tinker Creek*, where a cosmic strife between harmony and disorder, between beauty and ugliness, is shown to be an inescapable element of our lived experience of nature—but also inevitable as we survey the common experience of humanity, apart from the experiment we have recently undertaken in the secularized West. It was for Confucius the need to restore on earth the ways of heaven through the recovery of human-heartedness, and for Lao Tzu the return to the way and virtue of the Tao that is both hidden and manifest in things. It was the Hindu quest to see through the veil of Maya, and the Buddhist prescription to relinquish the grasping and clinging that bring suffering into the world and sustain it. And we could go on at great length, mentioning the cosmic warfare apprehended by the ancient Persians, and the loss of a golden age assumed by the Greeks and many other peoples. Indeed, the Promethean humanism discussed earlier is itself a one-sided appropriation of this insight. This is cosmic misalignment, which nevertheless finds its locus in humanity—yet from whom its redemption must proceed, as it does even in Buddhism and its teaching of the Bodhisattva, whereby liberation from cosmic errancy and suffering must take place through the medium of humanity. But if this is correct, then there is an even deeper truth to anthropocentrism, one having nothing to do with self-interest, enlightened or otherwise. As understood in the Byzantine tradition, humanity was created as a cosmic priesthood, the living link between God and cosmos, positioned to consecrate and celebrate the nature that surrounds us: in Bulgakov's terms, the "ontological center" of nature—"the unifying center of the world in the eternal harmony and beauty of the cosmos [as it was] created by God."[40] And it is precisely *because of* this cosmic role that the fall of humanity must at the same time necessarily be a fall for nature as well. Destined to serve as the medium through which creation itself was to become deified, fallen humanity becomes instead the bane of nature, its scourge and oppressor. Thus Edenic nature, in contrast, represents the normative "*anamnēsis* of another [mode of] being, similar to the golden dreams of childhood and most accessible to childhood . . . [proffering] distinct, palpable revelations of the world's sophianicity in our soul." Accordingly, it serves as "a *preparation* for what [lies] hidden in the recesses of all of natural being . . . [as] a sort of eschatology of natural being."[41]

But why does Bulgakov propose *art* as the paradigm of the Edenic economy, as the vehicle to realize this eschatology? Why not, instead, science—and especially ecology—as environmentalists tacitly presuppose? Science, he answers, can only study nature in its lifeless and mechanized, and thus its fallen, aspect. "Science throws a net of mechanism, imperceptible as are the threads of a spider web to a fly, over the entire world," and as a result, "nature as universal organism, *hen kai pan*, does not yield itself alive to science."[42]

It lacks entirely the means to see life as everywhere incipient in nature, to apprehend the sophic seeds awaiting germination, the spring day that Rilke says was waiting for us to celebrate it and sanctify it. Science can indeed trace the sophic footprints in *natura naturata*, but it is methodologically incapable of apprehending the life that left them behind, *natura naturans*. "*A scientific relation to the world*," writes Bulgakov in italics, "*is a relation to the world as mechanism*."[43]

As Kant had shown long before, the truth of things is inaccessible to discursive thought.[44] "Sophia, which establishes the ultimate connection of all things," Bulgakov maintains, "cannot be understood through science, which only understands nature's regularities and patterns."[45] Discursive knowledge fails here, because it is based on "the division of subject and object and on the disintegration of being," and thus it finds the truth of things surrounded by cloudlike mystery and inexpressibility. But this does consign us to either skepticism or blind belief. "Sophia can only be perceived by revelation"—not discursively, but noetically through "miraculous, intuitive ways independent of scientific cognition." It is revealed through religious myths and symbols; revealed, he says, in the brilliant intuitions of great philosophical genius; revealed aesthetically, in modes whereby "the infinite shines through the finite"; and revealed too in the inexpressible "mysteries of personal religious life."[46] In each of these, "*truth is a state of being*" in which we "become a living member of the divine Sophia . . . become transparent and sophic" ourselves, and thus "Sophia—that sun which shines and warms us while remaining invisible—emerges from the clouds and openly stands in the middle of the sky."[47] Bulgakov himself clearly seems to have experienced such noetic insights, as indicated by his autobiographical account of a twilight vision of the Caucasus Mountains from the Russian steppes—when, as he states it, "the first day of creation shone before my eyes," and liberated him from the dark dreams of Marxist materialism and "the dull pain of seeing nature as a lifeless desert and of treating its surface beauty as a deceptive mask."[48]

Bulgakov's *Philosophy of Economy* is a long and complex book; we have reviewed only some of its main features, and this in a summary fashion. Moreover, what was published in 1912 was only volume one (*The World as Household*) devoted to the ontology of economic activity. A second, projected volume would have delivered his economic axiology as a philosophy of history and culture, showing how a sophianic economy can be seen emerging in both human history and in culture. Since it never appeared, the content of Volume Two must be extrapolated from his later, theological writings—the task of another book, especially given the developments in the twentieth century that would need to be integrated into Bulgakov's perspective. Developments such as accelerating globalization, the rise of the environmental movement, and the steadily increasing interest of traditional religious institutions in integrating environmental perspectives into their theological outlook would all reinforce key concepts in Bulgakov, such as the ontological unity of humanity, and the hope for a progressive realization of sophianic economy.

But even what is developed in *The World as Household* provides powerful tools for rethinking environmental issues for it delivers a philosophical means for understanding environmentalism as an important mode of the great charge of humanity, and this is no less than—to use the biblical terminology of William Blake—a building of the New Jerusalem on earth, requiring no less than the resurrection and transfiguration of nature. Championing the nature revealed by poetry and imagination over the lifeless, mechanized nature of Newton and modern science—which he sees as powerless to check the rise of "dark satanic mills"—Blake ends his poem with this stanza:

> I will not cease from mental fight,
> Nor shall my sword sleep in my hand
> Till we have built Jerusalem
> In England's green and pleasant land.

To build the New Jerusalem, the earth as the Holy City, requires mental, spiritual, noetic "fight"—deployed through the sword of imagination, scimitar of the Edenic economy, and vanquisher of lifeless matter and dissonant disorder—to restore "England's green and pleasant land." But whence the criterion for this restoration, the touchstone for distinguishing the Edenic from the Satanic economy? The hallmark is the divine Sophia, the Eternal Logos as it permeates nature, *natura naturans* as it shows itself in what is beautiful, and consequently finds resonance within the human heart. If Bulgakov is right, this is not something vague, something secondary or merely subjective, but the very grounds for the interaction with nature, that makes us mediators between the finite and eternal, the visible and the invisible. And it inaugurates a far more powerful environmental vision than those of the warring schools of environmental philosophy—anthropocentrism and biocentrism, social constructivism and positivism. It allows us to understand why sound environmental restoration is not the fabrication of something artificial, but the re-vitalization of cosmic Sophia. As it is manifest in the monastic life of the Eastern Church, it even suggests to us that the call to meatless lifestyles may advance the environmental project not primarily in a utilitarian manner, i.e., through its causal consequences, by bringing us closer to the Edenic economy from which we have ontologically fallen. At the same time, it provides a means for understanding the domestication of animals not as only as a degradation (which, of course, in some cases it is) but in its most salutary forms, as an elevation of the pre-human into the higher modes toward which it inclines, and with which it instinctively cooperates, while at the same time restoring to us an animal identity that—as Paul Sheperd has argued—we ignore at our own peril. It allows us to subvert the destructive dichotomy between supposedly pristine wilderness and human habitation—as if human hands were necessarily contaminants—and allows us to understand Rene Dubos's controversial claim (in *The Wooing of the Earth*) that human activity can sometimes so much represent the tender hand of a householder in love with the land, that nature can actually be made *better* than it was before. It answers the questions of why the disorder and aberration entailed by catastrophic climate change, by genetically modified plant and animal species, and by the extinction of those species that were first articulated as individual notes of the great Song of the Earth—why all of these disruptions are sophic abominations, and not, as some would now argue, just as "natural" as anything that has come before. Seeing nature as inherently sophianic—not just empirically, but in the possibilities toward which it inclines—allows for a deep understanding of why technologies friendly to natural patterns are intrinsically better than the violent technologies of Promethean humanism; why natural fabrics, materials, foods are important to an environmental vision that seeks the Edenic; and why the natural species that express the sophic genius of place are always preferable to exotics, even though the latter are, in a secondary sense, just as "natural."

Bulgakov's writings—both earlier and later—merit sustained consideration from environmental philosophers. But more important yet, the thought of a normative depth to nature—accessible first in aesthetic terms, but requiring a religious, spiritual, mystical recursion to be rendered even minimally intelligible, a vision of what Bulgakov calls *sophianic* nature—has now become indispensable for serious thinking about the natural environment. It is, I conclude, a need become urgent and apparent, both in the literature of environmental thought, as well as in the events and processes that surround us, and threaten to overwhelm us.

CHAPTER 6

The Iconic Earth

Nature Godly and Beautiful

The silence of the earth seemed to merge with the silence of the heavens, the mystery of the earth touched the mystery of the stars . . . Alyosha stood gazing and suddenly, as if he had been cut down, threw himself to the earth.

He did not know why he was embracing it, he did not try to understand why he longed so irresistibly to kiss it, to kiss all of it, but he was kissing it, weeping, sobbing, and watering it with his tears, and he vowed ecstatically to love it, to love it unto ages of ages.

—*Fyodor Dostoevsky,* The Brothers Karamazov[1]

An elder was once asked, "What is a compassionate heart?" He replied:

"It is a heart on fire for the whole of creation, for humanity, for the birds, for the animals, for demons and for all that exists. At the recollection and at the sight of them such a person's eyes overflow with tears owing to the vehemence of the compassion which grips his heart; as a result of his deep mercy his heart shrinks and cannot bear to hear or look on any injury or the slightest suffering of anything in creation.

"This is why he constantly offers up prayers full of tears, even for the irrational animals and for the enemies of truth, even for those who harm him, so that they may be protected and find mercy.

"He even prays for the reptiles as a result of the great compassion which is poured out beyond measure—after the likeness of God—in his heart."

—*St. Isaac of Syria*[2]

Hidden Patency: The Omission of the Obvious

Environmental philosophy is relatively young, its features still unformed and undecided. Initially, it took shape as an environmental ethic with a straightforward orientation toward justifying practical imperatives, and often this meant little more than extending the boundaries of existing moral theories to include the natural environment. At the same time, it rarely questioned whether there were grounds for right action lying outside the properly ethical—that is, grounds based not on moral obligation, but on appreciation and reverence—i.e., based not on the ethical alone, but also the aesthetic and theological, and invoking not simply or even primarily justice, but the beautiful and holy as well.[3]

This early predilection of environmental philosophy was influenced by a tacit assumption that the discourse of duty and moral imperative would constitute the most effective response to the perceived urgency of environmental crisis, thereby allowing aesthetic and theological factors to be largely neglected. Yet in hindsight, this orientation seems peculiar and unfortunate for several reasons. First, while it is counter-intuitive to speak of inanimate or marginally sentient being as possessing rights and accruing obligations, it is commonplace to speak of those same domains in terms whose aesthetic and spiritual charge is quite explicit: to say, for example, that a natural setting is "lovely" or "inspiring." And even if we do end up judging that a land or landscape (say, one that has been clearcut, strip-mined, or strip-malled) has been "wronged," this will most likely be derivative from a more primary perception that it has been disfigured (made ugly) or desecrated (profaned). Why, then, have the aesthetic and theological been so consistently ignored in environmental philosophy, while the ethical has been singularly valorized? This neglect is made even more paradoxical by the fact that after thirty years of work in environmental ethics, hardly any philosophers outside a small circle, and virtually no one outside the academy, takes seriously the notion that "rocks have rights." That is, ethical arguments have, in fact, not only failed to serve as a bulwark against pollution and pointless development (their original justification), but they have hardly been persuasive at all.

This discounting of aesthetic and theological issues in environmental philosophy has not, of course, been total.[4] But the limited work published so far in these areas has been largely preliminary and tentative. Its paucity relative to the extensive literature published in environmental ethics over the past three decades is puzzling, especially in view of the many indications of the need for such work. For example, Max Oelschlaeger, in one of the few books devoted to considering the role of the sacred in environmental philosophy, concludes that "religious discourse, and especially [that] of the biblical tradition, is our best chance to escape the belly of the whale," i.e., "ecological catastrophe."[5] And key environmental writers consistently make the same point, even if this is more often implied than stated. Examining those writers in whom an exemplary relation to the natural environment has most often been sought (both predecessors such as Thoreau and Emerson and Muir and contemporaries such as Annie Dillard, Edward Abbey, Gary Snyder, and Barry Lopez) we find that they speak a great deal about its beauty and about its sacred character, while making little use of the discourse of rights and duties:

> Nature writing in America has always been religious or quasi-religious. All the important studies on the subgenre conclude that nature writing is "in the end concerned not only with fact but with fundamental *spiritual* and *aesthetic* truth." That is true of essays by Thoreau, John Muir, John Burroughs, Aldo Leopold, Edwin Teale, and Joseph Wood Krutch, whose works represent more than a century of American nature writing. And Edward Abbey's work is infused with spiritual impulse, as he engages "Mystery."[6]

This aesthetico-theological emphasis is also broadly represented in contemporary philosophy, and the two thinkers most widely cited as providing elements for a comprehensive environmental philosophy (Heidegger and Whitehead) both articulate their understandings of nature and the earth within a *topos* where the vectors of the sacred and the beautiful intersect. This is markedly so with Heidegger, who consistently framed his thinking regarding the earth with discussions of the gods and of the poetic, both in his writings on Hölderlin (where nature is seen as "divinely beautiful") and in his expositions of a poetically inhabited fourfold of earth and sky, gods and mortals. And if all this is does not make the point persuasively enough, we can also note that sacred and "mythopoetic" cultures have long lived sustainably upon the earth, while the only secular culture that the world has known (as Camus reminds us, the only one to have turned its back on nature, "ashamed" of the beauty of the earth, in order to immerse itself in history) has managed to devastate the earth in just a few hundred years.[7] Yet despite all these indicators, and despite the now obvious limitations of environmental ethics, environmental philosophers have been nearly silent on the aesthetic and theological dimensions of environmental issues. Why, then, has this omission of the obvious taken place?

Aesthetico-Theology and Onto-Theology

> Thus all religion would be poetic in its essence.
>
> *Friedrich Hölderlin*[8]

> How is one to understand this holy, this beautiful aspect of creation? . . . What is it metaphysically?
>
> *Pavel Florensky,* The Pillar and Ground of the Truth[9]

> Whoever wants to become a Christian must first become a poet.
>
> *Elder Porphyrios,* Wounded by Love[10]

The exclusion of the aesthetico-theological element in environmental philosophy is closely connected to the course of modern philosophy as a whole. One might, perhaps, begin to address it through the early *Aufhebung* of nature as mere immediacy in the philosophy of Hegel, or perhaps with Hegel's view—converse to that of the ancients, as well as of traditional peoples everywhere, and cited with approval by Gadamer—that "natural beauty is a reflection of artistic beauty."[11] Kant is important here, too, for depriving beauty of ontological standing outside affectivity and subjectivity, a step radicalized by Nietzsche's understanding of beauty as salutary illusion. We might also look to Kierkegaard, who rejects metaphysical inquiry altogether, and who places the aesthetic sphere beneath the ethical and the religious, thereby setting up a kind of ethical (and, perhaps, properly Protestant) quarantine between the aesthetic and the religious.[12] In each case, we find a turn away from metaphysics toward subjectivity, each of them reflecting philosophy's attempt during the past two centuries to retreat from metaphysics. This has created serious difficulties for understanding the aesthetico-theological dimensions of environmental problems, for as soon as we begin to grant the beauty of environment ontological standing outside subjectivity, it becomes difficult to avoid discussing divine transcendence and immanence—themes that are thoroughly theological and that have at the same time been bound up in Western thought with metaphysics.

What is discreditable about metaphysics? For Anglo-analytic thought, the problem is just that it is ultimately nonsense, and not a very interesting sort of nonsense at that. But the Continental rejection of metaphysics has never been so glib, always possessing some of the pathos of Nietzsche's tormented lantern bearer. Its complaint about metaphysics, beginning with Kierkegaard, was less that metaphysics erred by thinking about certain kinds of matters deemed frivolous or undecidable, but that it proceeded with insufficient earnestness—posed its questions, not in a way that made us take them too seriously (as analytic, and now some deconstructionist, philosophers seem to say) but rather one that was not serious enough. Nietzsche, parallel to Kierkegaard, sees metaphysics as simultaneously weak and enervating because it has constructed a pale and phantom world, a world lacking vigor and resistance. This non-dismissive, yet no less definitive, rejection is best epitomized by Heidegger's claim that metaphysics is onto-theology.

What, then, is wrong with onto-theology? For Heidegger, the problem is not on the left side of the contraction. The shortcoming lies not in posing ontological questions, but in posing them ontically, i.e., not radically enough, by posing the question of being in terms of the question of the highest entity. Nor does the objection rest on the other side of the conjunction. The problem here consists not in speaking about God, not in theology per se, but in speaking about a made-up God, a God who doesn't place demands on us, a mere metaphysical *causa sui* before whom we could never fall to our knees in awe.[13] Onto-theology is problematic because it has substituted for the "divine God" the "god of philosophy," an ontic god, a highest being, whose primary role is to regulate and legitimate our understanding of an ontic realm: nature as *ens creatum*. Especially important for environmental philosophy is Heidegger's claim that this onto-theological concept of God as *prima causa* has not only denigrated (and indeed, blasphemed) the divine God, but degraded and "de-natured" nature as well, "dis-enchanted" (*Entzauberung*) and even "de-deified" (*Entgotterung*) nature, freeing it for the new "enchantment" (*Verzauberung*) of technology.[14] If what is wrong with metaphysics is onto-theology, and if what is wrong with onto-theology is this understanding of the relation of God and nature, then at the same time it seems clear that what is discreditable about metaphysics is epitomized in the medieval scholasticism where these concepts come of age and assume a predominance that persists throughout modernity.

How, then, is it possible to speak about God and nature and the divine beauty of nature outside of onto-theology, i.e., without either obscuring "the divine God" or "de-deifying" nature? Heidegger generally finds an answer in the "saying" that characterizes poetry, and in the same essay where the concept of onto-theology is first presented, theology as "*theologos, theologia*"—as the "mytho-poetic saying [*Sagen*] concerning the gods"—is explicitly *excluded* from his critique and even offered in contrast.[15] It should not surprise us, then, that in essays and lecture courses spanning more than three decades, Heidegger follows the poetic discourse of Hölderlin concerning "divinely beautiful" nature. And indeed, Heidegger would seem to lend important support here for the hope that environmental philosophy could embrace the aesthetico-theological element without at the same time embracing onto-theology. But what must surprise us is the persuasiveness of an essay by Michel Haar, which argues that Heidegger, who has come to the poet to learn how to speak of divinely beautiful nature, in fact follows him rather poorly precisely in relation to Hölderlin's *theologos*, his "mytho-poetic saying concerning the gods." Haar discusses at length ways in which Heidegger seems to obscure and distort (a) those very elements of Hölderlin's poetry that portray God personalistically, and thus are consonant with Jewish and Christian experience, and (b) those suggesting that the difficulty facing the poet (and hence those would-be "theologians" who look toward the poet for help in learning how to speak) is not primarily the "default of God" and the "flight of the gods" at all, but simply "the extreme *difficulty* of celebrating them, the difficulty of poetic naming," i.e., the inherent difficulty, which is not an "impossibility," of aesthetico-theological discourse.[16] Haar's conclusions have dramatic implications. They suggest, for example, that Heidegger's interpretations of Hölderlin draw us forcefully *away from* the very possibility that religious factors could help matters here, since the faith-traditions which (in William James's language) are "living options" for most in the West are rendered inarticulate, and since God is seen to have defaulted anyway. And they draw us *toward* silence and a kind of pensive paralysis where we can only speak elliptically about the flight of the gods or else reflect apocalyptically on the conditions for the possibility of "*der letzte Gott.*"

Much has been made recently of Heidegger's own religious journey, from the biographical research of Kisiel, Van Buren, and Ott to the deconstructive studies by Caputo. Derrida, too, has spoken of Heidegger's inability "to stop either settling accounts with Christianity or distancing himself from it."[17] Perhaps the aversion Heidegger developed to the scholastic philosophy in which he had once been immersed prevented him from hearing the suggestion in "poetic saying" of more favorable theological alternatives? Perhaps it is even possible that Heidegger's emphatic rejection of medieval scholasticism colored retroactively his reading of classical metaphysics, for the "onto-theological constitution" of early Greek philosophy is by no means so evident as it is in Aquinas or Descartes or Hegel? Perhaps onto-theology, and thus what is discreditable about metaphysics, takes shape definitively only within medieval scholasticism—and does so with such force that it generally blinds us not only to the character of earlier metaphysics, but also to the same spiritual tradition that it attempts to render philosophical, yet at the same time serves to distort? Perhaps something occurs in the medieval West that amounts to a rupture with the past and the beginning of something new and eventually "modern," and whose elaboration continues to discredit philosophical talk about God and about beauty even today?

Perhaps, too, this break is connected to the attitudes that have led to our current environmental problems?

Technē East and West

And therefore I have sailed the seas and come
to the holy city of Byzantium.

W. B. Yeats, "Sailing to Byzantium"

The last of these questions, at least, is answered positively in what is easily the most influential essay concerning the relation between religion and environmental crisis. In "The Historic Roots of Our Ecologic Crisis," the historian Lynn White Jr., with no apparent influence from Heidegger or G. F. Jünger or Ellul, documents the rise of new technologies during the Western Middle Ages in language that resonates with virtually the same imagery as Heidegger's technology critique. Wind power was "harnessed" for the first time, and water power was extended far beyond the milling of grain to power the sawing of timbers and pump the bellows of blast furnaces.

The old "scratch plow" was abandoned in favor of a new kind of plow built like a gigantic knife blade to "attack" the earth with "ruthlessness" and "violence," while at the same time requiring plow-teams of eight oxen, and thus entirely new forms of social organization. "Man's relation to the soil," he notes, "was profoundly changed." Even the imagery on calendars went from traditional personifications of the months to the new Frankish calendar style, which illustrated "men coercing the world about them—plowing, harvesting, chopping trees, butchering pigs."[18] By the middle of the thirteenth century, as he had stated it several years earlier in *Medieval Technology and Social Change*, many of the most "active minds" in Western Europe were "coming to think of the cosmos as a vast reservoir of energies to be tapped and used according to human intentions. They were power conscious to the point of fantasy."[19] White, in turn, connects these new technologies, which arose centuries before modern natural science, with changes in Christian religious sensibility. Most important of these changes, he argues, was a new emphasis on the transcendence of God. But this further entailed, he continues, a new understanding of humanity, made in God's image, as likewise fundamentally transcendent to nature. And when this view of ourselves as essentially outside nature is combined with the divine charge recounted in Genesis to have dominion over nature, humanity would appear to have been granted a license for the technological domination of the natural environment.

Many have seized simplistically upon White's widely read article as grounds for rejecting Christianity altogether in the name of ecology, overlooking his conclusion that "since the roots of our trouble are so largely religious, the remedy must also be essentially religious."[20] But even more widely ignored is White's emphatic qualification that his critique refers specifically to the Christianity that arises in Western Europe during the Middle Ages, i.e., to the very modality of Christianity that Heidegger accuses of de-naturing, disenchanting, and de-deifying nature. White explicitly contrasts Latin Christianity with "the Greek East, a highly civilized realm of equal Christian devotion, [which] seems to have produced no marked technological innovation after the late 7th century," and whose very different "tonality of piety and thought" saw sin not as "moral evil" like the West, but as "intellectual blindness" in need of illumination, making it fundamentally contemplative rather than "voluntaristic." Even more important for us, however, is his characterization of the view of nature held by Byzantine Christianity (and by the tradition which it preserved) as being essentially aesthetic as well:

> In the early Church, and always in the Greek East, nature was conceived primarily as a symbolic system through which God speaks to men. . . . This view of nature was *essentially artistic rather than scientific.* While Byzantium preserved and copied great numbers of ancient Greek scientific texts, science as we conceive it could scarcely flourish in such an ambiance. However in the Latin West by the early 13th century, natural theology was following a very different bent. It was ceasing to be the *decoding of the physical symbols of God's communication with man* and was becoming *the effort to understand God's mind by discovering how his creation operates.*[21]

The last sentence of this citation is especially notable because the second clause ("the effort to understand God's mind by discovering how his creation operates") serves to summarize rather well Heidegger's charge against the medieval Latin doctrine of nature as *ens creatum,* while the first clause ("the decoding of the physical symbols of God's communication with man") is a fitting characterization of the task assumed by Hölderlin's theological verse. (In *Griechenland,* he writes "Everyday but marvelous, for the sake of men, God has put on a garment.")[22] At the same time, this characterization of "physical symbols of God's communication to man" describes what in Byzantine Christianity is called an "icon," even though it refers not to icons made by artists but to nature itself as iconic. And as I will maintain later in this chapter, the aesthetic of Eastern Christianity draws upon a theology maintaining that icons, godly images, can be made by artists only because the earth itself *is* fundamentally iconic. But it is important first to consider more generally the difference between the sacred art of the Greek East (and all of Christendom through the early Middle Ages) and the "religious" art that began to develop in the Latin West, parallel to the technological inception documented by White.

It takes some effort to appreciate the depth of this divergence, for it requires us to set aside the conventional narratives, born of Western ethnocentrism and Enlightenment triumphalism, in which Europe awakens from the crudity and barbarism of the "Dark Ages," first through scholastic philosophy and its culmination (and perhaps *reductio ad absurdam*) in the *via moderna* of Ockham, and subsequently through a re-naissance of classical humanism and worldliness, shining the light of realism and reason upon a natural world now freed from the other-worldly shadows of the divine and prepared for scientific scrutiny. We need to recall, on the contrary, that the decline of Rome and of Western Europe that began in the fourth century coincided with the ascent of Constantinople—Old Byzantium and New Rome—cultural center and capital of the Empire until its fall to the Ottoman Empire in the fifteenth century, and which prior to this underwent neither a "dark" nor a "middle" age. Thus, while much of the sacred art of Western Europe in the early Middle Ages may indeed have been crude and barbarous, this was simply because the West had become a backwater, weakened by invasion and conquest, and cut off culturally and linguistically from the Greek East where sacred art continued to develop and mature over the period of a thousand years into forms of unrivaled subtlety and depth.

But at the same time it is important to note that however rustic it may often have been, the sacred art of Western Europe throughout the early Middle Ages still employed much the same theological aesthetic as the sacred art of the Byzantine Empire. Both murals in Constantinople and mosaics in Ravenna, both the Coptic crucifix and the Celtic kell seek to make manifest the invisible within the visible, to show the eternal and uncreated embodied in the created order, and thus to show matter infused with divine light—nature speaking a language not of mathematics but of divine mysteries. At the same time, both are contiguous with the sacred art of extra-modern humanity as a whole, with Navaho sand paintings and the arrangement of stones in a Kyoto temple rock-garden, with illustrations of Krishna in pages of the Mahabharata and of the Buddha in the murals at Ajanta. Yet all these are rendered untenable (or re-cast as merely superstitious) by the scholastic notion of a "nature" that is subsistent and intelligible apart from what is now *ipso facto* "supernature" (Aquinas)—not to mention a natural world for which all that can be known must be gathered through sensation alone (Ockham). What modernist revisionism sees as a "freeing" of nature is just as much a severing of creation from the uncreated, an uprooting of the visible from the invisible.

In his monumental history of medieval art, French historian Georges Duby shows that through the eleventh century—the time of the Great Schism between the Eastern Orthodox and Roman Catholic Churches—art in the Latin West still proceeded from the same premises that continued to prevail in the Byzantine East: "a thirst for God—that is to say for mystery," an orientation toward seeing the inapprehensible in the perceptible, and thus a recourse to "the rhythms whereby the world is attuned to the breath of God."[23] In such a world, "like music and the liturgy, architecture and the visual arts had the nature of an *initiation*."[24] But by the thirteenth century, he notes a very different world characterized by a "dazzling" sense of the new, a sense of "modernity," a marked increase in material wealth, and a powerful trend toward the popular and the secular.[25] And connected to the new teachings of scholastic philosophy, he finds both a rapid increase in the development of the exact sciences after 1280, and a shift whereby "ecclesiastical art gradually moved from the essence of what it portrayed to its actual appearance," culminating in the work of Giotto that Boccaccio praised for "reproducing" and "imitating" nature so exactly that "the painted" was often taken for "the real."[26] And concerning the treatment of space developed by the Florentine painters of the thirteenth and early fourteenth centuries, Duby observes that far from a space of "mystic illumination," it was rather "the space of common sense, public policy, and hard work. It was intended for men who built town clocks to divide up the day into hours, men who wanted to see where they were going and how much they were making. . . . [for] men who knew the price of a sack of corn or a bale of wool."[27]

But doesn't the renowned emergence of such desacralized "realism" in the later Middle Ages, "opening the road towards an uncompromising secularization," represent an intellectual celebration of this-worldly nature, paving the way for the Renaissance?[28] (It should be remembered here that both Cimabue—who appears in the *Purgatorio*—and Giotto were contemporaries of Dante.) Or does it even more significantly represent a practical celebration of new possibilities for the conquest of nature, proceeding from the scholastic appropriation of Aristotelian physics that had resulted in "an abandoned world, a world cut off from the transcendent," and that is thereby prepared to be the neutral material for the mastery of *either* the artist or the engineer?[29]

Duby documents one of the earliest graphic representations of nature in this new mode. He notes that the first books of natural history in the modern sense (itself the precursor to the science of ecology) were not scholarly treatises but hunting manuals, detailing "a form of conquest [that] enabled men in their daily lives to seize and subjugate the humbler creation." So far, this agrees with White's observations on the rise of new technologies and corresponding attitudes toward nature. But Duby goes on to note that this same material found an additional graphic form of expression: "People were delighted when animals or plants they had observed in the hunting-field reappeared along the margins of their Psalters or breviaries."[30] And in the worldly rendering of an ivy vine or a stag adorning the liturgical text, we find an inconspicuous eruption into the sacred realm, of nature not as divine communication but as presented through the lexicon of human conquest and domination. Soon this mode of representation moves from the margins onto the center of the page. Yet before this new aesthetic sensibility began to make its way into the brightly colored breviary, it had long been emerging in the perceptual realm itself.

Umberto Eco has shown a two-stage transition in the medieval aesthetics of nature, which parallels the developments in technology and art documented by White and Duby. The first stage of alteration was a transition from the sense of what Eco calls "metaphysical symbolism" in nature to an attunement toward what he calls "cosmic allegory." The former refers to the "philosophical habit," inherited from Greek antiquity, and having analogues among traditional peoples everywhere, of "discerning the hand of God in the beauty of the world," of seeing nature as "a great theophany" in which "the face of eternity shines through the things of the earth."[31] That is, in the early middle ages of the Latin West, nature was seen just as it continued to be seen in the Byzantine East, precisely as iconic. The second, later stage of cosmic allegory sees nature not as a theophanous manifestation but as discursive representation of a codified "supernatural," that now needs to be interpreted by means of an increasingly conventional system of tropes, types, and figures. But as growing skepticism arose concerning the "allegory of nature," and as the proper realm of allegory came to be seen as scripture and art rather than the cosmos, a third aesthetic arose. Initiated by the master of scholasticism, Thomas Aquinas, this new aesthetic was, in Eco's words, more "scientific," more "humanistic," more "rational"—a "functionalist theory of beauty" tending "to identify the beautiful and the useful."[32] Once again, this development can be seen either as completed, or as driven to bankruptcy, by Ockham, but at any rate, the result was a conceptual framework in which "the problem of the transcendental status of beauty. . . can scarcely [even] be posed," and which indeed "made a metaphysical beauty impossible."[33]

By the time of the Renaissance, the divergence—both in artistic representation, and in the aesthetic apperception of nature—between East and West had become as complete as the ecclesiastical divergence between Latin and Byzantine Christendom, which took 1054 as its date, but which needed much longer to actually unfold. The religious painting of the Renaissance, which has so much been valorized that it has become exemplary in Western Europe, is no longer sacred art at all, i.e., distinctly sacred art with its own indigenous style and sensibilities, but simply secular painting that happens to have a religious theme. All the techniques of secular art (such as realist exactitude, the use of actual human models rather than traditional imagery, and newly mathematicized techniques of linear perspective) are employed to illustrate themes that happen to be religious, but in precisely the same manner that secular themes would be portrayed. The fifteenth-century Florentine painter of religious art is a secular master of worldly techniques (quite possibly a figure of wealth and power) whose aesthetic commitment to the sacred is primarily professional—and who stands in sharp contrast to his contemporary in Constantinople or Mistra, on the eve of the Ottoman conquest, who is committed to remaining anonymous, is unlikely to be wealthy, is almost certainly a monk or nun, and who is still expected to engage in the traditional period of prayer, fasting, and meditation before beginning an icon, which will be done using traditional materials and techniques to render sacred realities in ways carefully preserved within the living memory of an ancient faith-community.[34]

The Nature of the Icon

> From the visible image, the spirit launches itself toward the divine. It is not the object (the material icon) which is venerated but the Beauty which . . . the icon transmits mysteriously . . . the unmistakable lightning flashes of divine Beauty.
>
> *Joseph of Volokolamsk*[35]

> Gaze, then, upon every icon in simplicity of heart, for any doubt proceeds from the Devil in order to divert you from heartfelt prayer. Say to him: *the whole earth is holy*; the power of my Lord and of His most pure Mother—the Queen of the whole world—*is in every place* . . .
>
> *St. John of Kronstadt*[36]

The warm glow of golden luminescence that forms the element of every icon could not be more different from the optically corrected, mercantile light of Renaissance religious painting, a difference as great as that between the insistence of St. Gregory Palamas, the Athonite, upon the appearance of the uncreated light in nature—its appearance to Moses on Mount Sinai, to the three disciples at the Transfiguration on Mount Tabor, to Palamas's monastic contemporaries on Mount Athos chanting prayers as the golden sun sinks into the surrounding Aegean Sea—and the equally emphatic insistence of his English contemporary William of Ockham, set forth under the wan northern light of Oxford, that when it comes to nature, what your eyeballs see is precisely what you get.[37] Viewed from the East, the enlightenment of the *via moderna* is perhaps similar to the twilight hour when the window pane becomes illumined from inside as the sunlight fades and the lights are switched on: The pane itself becomes reflective of the artificial light, and what had been a window becomes an object unto itself. Nature, that is, becomes different. Under such conditions, it is no longer iconic. It is no longer sacred, except indirectly through the discursive reasoning of onto-theology, which by a fortuitous coincidence turns out to be even more useful for its mastery.

Icon, *eikōn*, names not only a sacred art, a window and bridge to the transcendent, but also a conceptual portal to the great and pivotal dyadic concern of philosophy: image/original, other/same, appearance/reality, temporal/eternal, immanent/transcendent. But is Western onto-theology the only rightful heir to this problematic? In the Byzantine East this issue was generally regarded as having been decisively resolved—not through philosophical effort, but by divine initiative, i.e., through the Incarnation of the eternal *Logos*, although the details of the resolution gave rise to controversy and ecumenical councils that lasted for half a millennium. Because of the Incarnation, they maintained, the earth has now begun to have restored to it its paradisiacal character, to regain its limpidity as divine image, and therefore the images created by artists could have this same character as well. But this new sacred art itself gave rise to yet more controversy, and yet another ecumenical council, during the iconoclastic crisis of the eighth and ninth centuries. And in the course of defending the making and veneration of icons, Byzantine theology was forced to clarify the iconic character of the earth.

Seen from within the usual Aristotelian distinction, Eastern Orthodox thinking is radically *practical*, providing a sharp contrast to the *theoretical* orientation of Latin scholasticism. It is concerned not with arriving at a set of correct propositions regarding God and nature, and even less with the onto-theological project of knowing things as they are known by God, but rather at mapping a pathway for the soul in search of healing (*katharsis*) and divine transformation (*theōsis*)—the mystical process that the West came to understand narrowly through the juridical notion of "justification." This practical orientation explains how, by abjuring a "theoretical" stance, Orthodox Christianity stood apart from the development of onto-theology in the West. It also explains the consternation of Latin visitors to Constantinople, who were amazed to come upon animated debates of seemingly subtle theological issues taking place in produce markets, for theological error was seen not as an intellectual concern for academics, but as the sort of thing through which one might get seriously lost. Finally, it explains the intensity of the iconoclastic controversy, which lasted more than a century: seen from one side, error entailed the sacrilege of idolatry, and from the other an equally sacrilegious contempt for creation. These were, in the East, by no means "theoretical" matters.[38]

That the iconodules, the defenders of the icon, prevailed is based at least in part on their success in arguing that Jewish and Christian scriptures alike, far from commending an iconoclastic contempt for matter as incapable of bearing divinity, continually show it as being filled with divine grace and activity. The great defender of iconic art, and hence of the iconic character of the earth, was St. John of Damascus in the eighth century, who based his argument for the sanctification of matter not just on strictly Incarnational premises but also on the history of revelation in the Old Testament as well, which narrates how the earth served as an icon of divinity in "the burning bush, the rock that gushed with water, the jar of manna, the fire from God that came down upon the altar"—all manifestations of what Palamas was to call God's uncreated energies in nature.[39] St. John of Damascus further emphasizes God's command to Moses to gather together earthly materials to serve as an iconic "glorification of matter":

> Whoever is of a generous heart, let him bring the Lord's offering: gold, silver, and bronze; blue and purple and scarlet stuff and fine twined linen; goat's hair, tanned ram's skins and goatskins; acacia wood, oil for the light, spices for the anointing oil and for the fragrant incense, and onyx stones and stones for setting.[40]

But although this iconic capacity of the earth had formerly been displayed episodically, it has now become generalized: because God has "deigned to inhabit" matter, indeed has "become matter," the earth itself has now become inherently venerable. Damascene continues: "Because of this I honor all remaining matter with reverence because God has filled it with His grace and power."[41] The veneration of icons is not idolatrous, as the iconoclasts charged, joining the Manichaeans and Gnostics in disparaging the visible and material. By showing contempt for the earthly, they disdain not only the goodness of the created earth in its first paradisiacal state, but also the partial restoration of this state in an earth whose very ontology has been altered.[42] Because the earth is now iconic in a radical and irrevocable sense, those who despise the earthly, simultaneously despise the divine, showing contempt for the original by despising its image.[43]

It is paradoxical that the iconoclasts were opposed most vigorously and most consistently by the same Latin West that was to later abandon the iconic tradition of sacred art for a secular realism, while Byzantium, where the iconoclastic heresy first arose, preserved and elaborated iconic art, along with the theology and liturgical practice that supported it. Liturgically, the veneration of icons has continued to this day to play a leading role in Orthodox worship, which dedicates the first Sunday of Lent to commemorating the "Triumph of Orthodoxy" in the victory over iconoclasm and the restoration of the icons. (The "Kontakion" for Matins of that day celebrates the Incarnation itself for having "restored the sullied image [*eikōn* of God in creation] to its ancient glory, filling it with the divine beauty," and thus making the sacred icon possible.)[44] Theologically, iconic practice found support not only in the overt defense of iconography by St. John of Damascus, but also in the more broadly cosmological thought of St. Maximos the Confessor (seventh century) and St. Gregory Palamas (fourteenth century).

Assimilating, while transforming, the vocabulary of Stoic philosophy, and its elaboration in the Platonic synthesis of the Jewish philosopher Philo, to the mystical vision of desert monasticism, Maximos propounds a natural contemplation (*theōria physikē*) of the *logoi* inherent in the natural world (*logos physeōs*). Unlike the *rationes* of St. Augustine, whose view became decisive in the West, the *logoi* of Maximos are not only universal, but singular to each created being, even while each *logos* is contained by the eternal *Logos* in which they find their unity, yet of which each serves as a unique image. Every entity, every shrub and stone, offers to the one who is spiritually prepared, its own glimpse of the "mysterious and deifying presence of Christ the Logos in the world."[45] Every being possesses as what is most its own, a *logos* which at the same time images the eternal Logos somewhat as (to use the example of a nineteenth century Russian representative of the same tradition) a face can be variously reflected in a myriad of tiny bubbles or drops of dew.[46] Nature, that is, is naturally iconic.

While Maximos maintained that natural beings are present in God, St. Gregory Palamas maintained that God can be not only present in nature, but can actually be experienced and known. Palamas bases this claim on a threefold set of distinctions: (1) between apophatic and kataphatic theology; (2) between the divine essence (*ousia*) and the divine energies (*energeiai*); and (3) between noetic intuition (*nous*) and discursive rationality (*dianoia*). He begins with the traditional claim, derived from the *Mystical Theology* of St. Dionysios the Areopagite, that the "negative" or apophatic theology, which emphasizes the utter incommensurability of God and creation (and hence, the complete inadequacy of our thoughts and words to comprehend the divine) is superior to the "positive" or kataphatic theology, which would make affirmative statements about the divine. Proceeding by way of negating all positive characterizations of the divine, theology properly ends in silence. Yet Palamas argues that this silence is not just a lack of speech but the very element of mystical experience, the precondition for the noetic vision (*nous*) of the divine. What surpasses discursive reason (*dianoia*) is not just the silence of negation, but the mystical vision which that silence makes possible, a conclusion allowing him to maintain that we can negate not only affirmation but negation as well because the divine life transcends both affirmation and negation. But his radicalizing of the apophatic does not entail his abandoning its central insight, and Palamas is insistent that the divine essence (*ousia*) is utterly unknowable and entirely unspeakable—thereby rendering onto-theology, and all speculative metaphysics, presumptuous and impossible. What is known of God, either noetically or discursively, is not the divine essence at all, but the divine activity or energies (*energeiai*), God as operative *within* nature, as self-revealing *in relation to* creation: "that which is seen through created things is not God's essence but his created energies."[47] Yet these divine energies, Palamas asserted in his debate with the Calabrian philosopher Barlaam, are nevertheless God (i.e., they are uncreated, just as the divine essence is uncreated). And with this, Palamas arrives at what in fact can be seen as a further elaboration of the *logoi* of Maximos: inhering in every entity, enabling it to be itself, is the uncreated energy of God, drawing it toward union with the divine even as it manifests the divine activity within the world. Nature, that is, is doubly iconic.

Taken together, Maximos and Palamas provide a sharply contrasting alternative to Western views of nature: not just to the understanding—gradually evolved in the Latin Middle Ages, but most fully articulated in Enlightenment deism—of nature as autonomous and self-contained, but also its presupposition in the scholastic view of nature as a realm intelligible apart from grace or "super-nature," and to which the latter would serve as a supplement. "The Eastern tradition," argues Lossky, "knows nothing of 'pure nature' to which grace is added as a supernatural gift. For it, there is no [purely] natural or 'normal' state . . . . 'Pure nature,' for Eastern theology, would thus be a philosophical fiction . . ."[48] Nature, rather, is permeated with divinity, abounds in divine energies, is upborn by divine *logoi*—yet it is not divine. This is neither a pantheism nor a panentheism. Nature so understood is not God, nor is it part of God, but rather divine creation to which the Creator "preserves an inner connection . . . through His uncreated energies": It is divine gift, self-expression, and image.[49] It is *iconic.*

Features of the Icon and of the Iconic Earth

> The world is one . . . for the *spiritual world* in its totality is manifested in the totality of the perceptible world, mystically expressed in symbolic pictures for those who have eyes to see. And the *perceptible world* in its entirety is secretly fathomable by the spiritual world in its entirety, when it has been simplified and amalgamated by means of the spiritual realities. The former is embodied in the latter through the realities; the latter in the former through the symbols. The operation of the two is one.
>
> *St. Maximos the Confessor,* Mystagogia[50]

> I have repeatedly been brought to a sudden state of awe by some gracile or savage movement of an animal, some odd wrapping of a tree's foliage by the wind, an unimpeded run of dew-laden prairie stretching to a horizon flat as a coin where

a pin-dot sun pales the dawn sky pink. I know these things are beyond intellection, that they are the vivid edges of a world that includes but also transcends the human world.

Barry Lopez, About This Life: Journeys on the Threshold of Memory[51]

The writings of Saints Maximos and Gregory Palamas are part of a vast library of mystical writings preserved by Eastern Christianity, and devoted to the ascetic (i.e., practical and experiential) theology that is characteristic of the traditional monasticism born in the deserts of Northern Africa and the Middle East. Yet unlike the Western Church, the East has not seen mystical experience as the province of specialists, but rather as the truest vocation, if not the norm, for everyone.[52] These writings, then, would be properly understood as a series of handbooks on attaining that end, as containing a theology that is oriented entirely to the practical and experiential, and that takes no interest in the speculative or theoretical for its own sake—handbooks devoted not to attaining an experience of God in the next life, but rather in the present life: for encountering the invisible within the visible, for using the silence of discursive reason as a springboard to spiritual vision. And if phenomenology in the broadest sense can be seen as radical empiricism, as being a disciplined approach not just to immediate experience but to what is usually concealed in immediate experience, then theology of this sort can be seen as a peculiar kind of phenomenology, an apophatic phenomenology. It would be a phenomenology whose *epochē* would be an *askēsis* from the "natural attitude" of conventional worldliness, and whose intentionality would be constituted by the soul's inner predilection toward invisible, noetic reality as it is manifest in the visible realm. But this is not just Eastern monasticism, nor is it merely Platonism, nor Romanticism, nor mythopoiesis. It is, of course, to some extent all of these things, but the central thesis of the present chapter is that it also constitutes the structure of the sensibility that is characteristic of those who have written most powerfully about the human relation to the natural environment, from Hesiod and Virgil to Barry Lopez and Annie Dillard; from the Psalms to Emerson, Thoreau, and Muir; from Lao Tzu to Scott Momaday.

To even modestly document this claim would require a separate and much larger study examining texts from many times and places. But it is possible here to indicate, at least in outline, how the main features of this iconic spirituality of Eastern Christianity can be found to lie at the heart of the experience of nature characteristic of those figures in the West to whom we commonly look for a normative vision. The possibility of an apophatic phenomenology, of an account of the visible transfigured by the invisible, although emerging in ascetic theology, would thus be tested through its ability to elucidate normative encounters with nature within what for the most part are overtly secular contexts. In the series of parallels presented in the following sections, the Eastern understanding of the icon, of its aesthetics as well as its numinous capacities, will be set alongside notable literary accounts of the earth, with the intent to show that similar structures of experience are at work in both.

The Materiality of the Icon/Earth

Icon Given the ancient abhorrence of idolatry that Christianity inherited from Judaism, the icon as an object of veneration has always retained, at first glance, a shocking character. Defenders against the iconoclasts argued that this is simply derivative from the shocking character of the Incarnation, that the eternal and invisible God should be born in flesh and blood, should have "accepted to be seen" (St. John of Damascus).[53] Yet the potential for abuse is inherent in the very materiality of the icon—in the idea of the holy as visibly presented in matter. Dufrenne attributes something parallel to this for all art precisely because it is addressed not to our understanding but to our *perception:* an "imperious presence" that derives from "the irresistible and magnificent presence of the sensuous," a "massive presence of the [aesthetic] object which almost does violence to us," even allowing us to "bestow the title of nature" upon the work of art.[54]

But the icon heightens this disconcerting "nakedness" of the sensuous that the aesthetic object puts before us (its "essential otherness which instrumentality masks") in a surprising way, by simultaneously employing this materiality itself to "clothe" the holy and the invisible.[55] It does not stop with "the sensuous appearing in its glory," but instead appropriates this brute character of the sensuous (this "incomprehensible" character of nature itself) *inversely* to draw us toward the non-sensuous.[56] There is no polemos or strife in the earthiness of the icon—earth is not set over against world here, as it should be, given the Heideggerian understanding—but rather the traces of the invisible are held gently, cradled in the visible. In the icon "matter is very much alive," comments Paul

Evdokimov, "but it is immobilized, so to speak, in contemplative quietness, so as to listen closely to the revelations."[57]

Earth Heidegger, of course, has made an immense contribution to this entire problematic by rethinking the philosophical category of "matter" as earth, and his work forms, in important ways, a starting point for this chapter, and for the book as a whole.[58] His understanding of the earth in terms of supportive self-withholding, of self-secluding sheltering, represents a decisive advance over the hylomorphic and atomistic concepts that have been dominant in Western philosophy. Less appreciated is the emphasis he places on the historical character of the earth, both in his elucidations of Hölderlin's poetry and in his autobiographical accounts in writings such as *Der Feldweg*. The earth, understood concretely and phenomenologically, is bound up with narrative and tradition, a feature to which we will return later in discussing the contextuality of the icon.

Nevertheless, the "earthiness" of the earth, its element of resistance to either narrative or conceptual determination, is its chief characteristic, thereby lending it an essential density—giving it an elemental opacity—and making possible its nurturing and supporting character. This tacit recalcitrance of the "earthy" is indicated by the manner in which mythic narratives are often tied to specific places on the earth, and thus are misunderstood when they are generalized, or else given a locus, Jungian style, in the psyche. In the accounts of notable nature writers, this earthiness is experienced as a kind of radical particularity, and often this is seen as making places and things precious and unrepeatable: this very maple leaf, the Hetch Hetchy Valley or Glen Canyon, which the earth will never again know. Eugene Hargrove has argued that aesthetically, "*existence* plays a more fundamental role in natural objects than in art objects," and hence the imperative to preserve natural beauty is even stronger than that to preserve great works of art.[59] Natural beauty has no "pre-existence in the imagination," as Hargrove puts it, or we might say, no concept of its own independent from its sheer existence, and its destruction is therefore more calamitous. But is it sheer existence that is most important here, or rather materiality—the fact that the earth itself has assumed this form, taken this shape? Does the wonder of the Yosemite Valley consist more in the fact that such a splendid place exists, or rather in the fact that it is the earth itself that shows forth this visage? The materiality of the icon is illuminating here: What is remarkable and ever surprising is that what is inherently invisible, what is sacred and holy, has here become visibly incarnate in and of this earth itself.

Thoreau was more attuned to earthiness or materiality than perhaps any other author, and he plays with imaginative inclinations aimed at intensifying his sense of the *irdisch*—for example, his expressed wishes to spend the day immersed to his neck in a swamp, or to devour a woodchuck alive. And in at least one case, he appears to have gotten more than he bargained for. Describing his ascent to the top of Mt. Ktaaden, an account in which the parallels to Moses on the holy ground of Mt. Sinai are unmistakable if unstated, he relates an experience of numinous materiality:

> Nature here was something savage and awful, though beautiful. I looked with awe at the ground I trod on, to see what the Powers had made there, the form and fashion and material of their work. This was that earth of which we have heard, made out of Chaos and Old Night. . . . . It was the fresh and natural surface of the planet Earth, as it was made for ever and ever . . . but here not even the surface had been scarred by man, but it was a specimen of what God saw fit to make this world.[60]

The evocations of the earth here as "creation" are striking and unmistakable, and coming from an author who was by no means inclined toward conventional theism, all the more noteworthy. They state eloquently what are perhaps the genuine roots of what, in its propositional articulation, became known in philosophy as the cosmological argument, an enterprise that is indicative of what happens to iconic vision when it becomes misunderstood through discursive rationality, and thereby points away from the very goal toward which it is supposed to lead. The iconic experience of the earth in its very earthliness is not just a starting point, but an end point as well.

And when Thoreau suddenly becomes aware that the startling materiality he finds on the holy mountain-top is the stuff of his own body, he concludes with ecstatic prose and a breathless account of what early traditions (Eastern and Western alike) once understood as "creatureliness"—and which is also, perhaps, better understood as an iconic experience of material embodiment than as the basis for an inferential argument regarding a causal sequence of events:

> I stand in awe of my body, this matter to which I am bound has become so strange to me. . . . What is this Titan that has

possession of me? Talk of mysteries!—Think of our life in nature,—daily to be shown matter, to come in contact with it,—rocks, trees, wind on our cheeks! the *solid* earth! the *actual* world! the common *sense! Contact! Contact! Who* are we? *where* are we?[61]

The Icon/Earth Is Non-Mimetic

Icon As does all sacred art, the icon seeks to present spiritual realities in a material medium (i.e., to present, within the visible, what is inherently invisible). And since what the icon presents is what is invisible and has always been invisible, it does not visibly re-present what has once before been visibly present itself. It is not representational or mimetic in the Platonic sense. It is not that the icon is imperfect in making visible what has already or otherwise been visible, but rather that it makes manifest that which itself is not visible and never has been visible. Ontologically, we could say that what is at work in the icon is not representation but something closer to presentation, less *Vorstellung* than *Darstellung*.

But aren't the figures of Christ and the saints necessarily representational? They are indeed in the religious art of the West, where the aim has become not epiphany (*epiphaneia, Epiphania*), a becoming present of the invisible, but *anamnēsis*, a recollection of past events.[62] And it indeed follows that if the latter were to be the aim, then the more accurate representation would indeed generate the more vivid "recollection." But the goal of the icon is not to render the ocular appearance of holy men and women at all, but rather their inner, spiritual reality. So while Giotto, Masaccio, Ducio, and Cimabue were introducing naturalistic innovations such as optical illusion, depth perspective, and chiaroscuro and initiating a realistic paradigm that would end up with the triumph of statuary (prohibited in sacred art of the East), the East itself continued to produce icons in the ancient fashion with results that thereby, with their "thin and elongated figures of extreme grace and elegance," seemed merely distorted from the new, Western point of view.[63] Likewise, eyes and ears continued to be altered from their naturalistic shape, precisely because the aim was still not to exhibit the painter's mastery of anatomical replication, but something quite different: to show how the senses of holy people are attuned to another, additional dimension, to show senses in the act of perceiving "what usually escapes man's perceptions, i.e. the perception of the spiritual world."[64] That is, the East continued its orientation toward the invisible and spiritual. Evdokimov maintains:

> The bodies drawn to underline their sveltness seem to float in the air or melt into the ethereal gold of the divine light; they lose all carnal character. The icon represents a world apart, renewed, in which persons with eternity written on their faces live freely together with the divine energies. These saints are energized by *epectasis*, the stretching out of a universe that dilates without limits in the heavenly spaces of the Kingdom.[65]

This is an art not of representation but of transfiguration, not of recollection but of presence to spiritual realities. And the latter is closely connected to the liturgical functions of the icon, for in the Orthodox liturgy it is believed that saints and prophets and angels, Christ and Theotokos, presented in and through the icons, do themselves become present and join together with clergy and parishioners in their worship and song.

It is possible to speak of the icon as symbolic, but only if *to symbolon* is taken in its ancient meaning of a coming together of two halves, in this case the meeting of the invisible and the visible. As opposed to the sign, whose relation to the signified is conventional and operative only within a system of signs, the symbol in this stronger sense "contains in itself the presence of what is symbolized."[66] "The icon is not a representation of events," says Archimandrite Vasileios, "it is Grace incarnate, a presence and an offering of life and holiness."[67]

Earth The beauty of nature, too, is non-representational in just this sense. If it is taken as a making-manifest, what it makes manifest is something inherently invisible, indefinable, and (ultimately even for the poet) ineffable.

In the first chapter of *Pilgrim at Tinker Creek*, a book that concerns itself especially with whether the beauty of nature can be trusted as offering a glimpse of transcendence, Annie Dillard lists a series of incidents in which this beauty was exhibited in an exceptionally noteworthy way. The last of these involves what to an unconcerned passer-by would appear to be no more than racing clouds on a windy day:

> The wind is terrific out of the west; the sun comes and goes. I can see the shadow on the field before me deepen uniformly

> and spread like a plague. Everything seems so dull I am amazed I can even distinguish objects. And suddenly the light runs across the land like a comber, and up the trees, and goes again in a wink: I think I've gone blind or died.
>
> It's the most beautiful day of the year. At four o'clock the eastern sky is a dead stratus black flecked with low white clouds. The sun in the west illuminates the ground, the mountains, and especially the bare branches of trees, so that everywhere silver trees cut into the black sky like a photographer's negative of a landscape. The air and the ground are dry; the mountains are going on and off like neon signs. Clouds slide east as if pulled from the horizon, like a tablecloth whipped off a table. The hemlocks by the barbed-wire fence are flinging themselves east as though their backs would break. Purple shadows are racing east; the wind makes me face east. . . .[68]

Earlier in her enumeration, Dillard has additionally listed the unexpected fall in flight of a mockingbird and the sight of countless sharks feeding in waves off the east coast of Florida, and the beauty that she finds in these three incidents, each chosen as exemplary, she characterizes as variously manifesting power, grace, extravagance, and mystery. Each of the latter, in turn, she connects with transcendence at play in nature, adding that she approaches them "with fear and trembling." But it would be absurd to claim that these incidents mimetically *represented* divine power, grace, extravagance, and mystery, as a studio portrait represents a person, or even that they were *illustrations* of these things such as we might find occurring in a didactic novel. Rather, they manifest these very things themselves, making them visible.

But at the same time, it would be just as erroneous to claim that the cloud-driven light-and-shadow show, the falling bird, the shark-disclosing waves were *examples* of divine grace, power, extravagance, and mystery in the same, straightforward way that the mockingbird was an example of a bird, the shark of a fish, and the sky of a meteorological condition. No doubt innumerable passers-by could have seen these same things while remaining unimpressed, just as the Rublev icon could look like merely a bad portrait. Iconic seeing is not correct seeing that could result from the rendering of additional information—the knowledge, for example, that some fish are really very large, allowing the ichthyologically benighted to see the shark as fish after all—but rather engaged seeing, inspired seeing, epiphanic seeing. The noetic seeing for which the icon calls, and the seeing of divine grace and power in nature that Dillard describes here, both result from a particular kind of engagement. Iconic seeing is transactional.

The Icon/Earth Is Transactional

Icon The icon is epiphanic only to the person who approaches it with piety and in the belief that it is holy. Nothing, then, could be farther from the kind of vision for which the icon calls than the "disinterested" seeing valorized in the aesthetics of Kant and Schopenhauer. Icons can, of course, be "appreciated" as examples of folk art, of high Byzantine culture, or of the "Treasures of the Czars," but then they are being seen from the outside, in a way that they were never intended to be viewed. The icon, instead, was made for a church, or a monastery, or for the icon corner of a home, where it will be venerated in the course of prayer and worship. It was, then, made for use within an interactive context, indeed a *trans*-active context, for worship and prayer are deeply transactional in their intentionality.[69]

The icon is understood within the context of a genuine reciprocity. That is, the act of looking is not itself a one-sided inspection. "The icon," states Kartsonis, "manifests its prototype alive and capable of presence, thought, will, action, interaction, and dialogue with the viewer."[70] "The Orthodox believer," state Onasch and Schnieper, "sees himself as observed by the holy persons portrayed in the icons, and only in a secondary sense is he or she an active observer."[71] Painting techniques employed by the iconographer contribute to this. Eyes are frequently painted so that they seem to be looking at us, no matter where we stand. And in contrast to the *optical perspective* that was developed in the Renaissance West—and which allows the viewer a comfortably clear, detached, and disinterested view—Eastern iconography has from antiquity employed an *inverse perspective,* in which the perspective lines come together in front of the icon, converging on the viewer, rather than retreating backwards toward a vanishing horizon, making the viewer an "outside observer." In this way, in encountering the icon "the observer becomes the observed; the person in the icon is really the one doing the looking."[72] St. John of Kronstadt encourages the pious to "look upon every icon as upon the saint himself or herself to whom we pray as living persons conversing with us, for they are as near, and still nearer, than the icons, if only we pray to them with faith and sincerity."[73] The icon can be epiphanic, in part because it is made precisely to engage and involve the viewer:

> "Time and nature are made new": worldly space is transfigured; [optical] perspective, which puts man in the position of an outside observer, no longer exists. The believer, the pilgrim, is a guest at the Wedding. He is inside, and sees the whole world from the inside.[74]

Earth Within the canon of nature writing, perhaps no characteristic is more consistently applicable than the serious engagement of the author with nature. This is often accompanied by the authors' observations that it is just this involvement, often presented explicitly as a quest or pilgrimage, and often following a withdrawal from the conventional world, that has enabled them to gain their revelations. They build cabins on ponds or creeks, or move into trailers in the desert; they spend days on their hands and knees observing up close, or paddling between icebergs in kayaks; but most of all they come to nature as seekers, looking for answers, for solace, for blessings: They come as suppliants. Annie Dillard is constantly posing questions, and indeed challenges, to God in the course of her encounters, and Muir scrambles through the woods like a starry-eyed lover on the heels of his beloved. Muir addresses nature, speaks to it, and listens to it speak, often in theistically charged language:

> When the storm began to abate, I dismounted and sauntered down through the calming woods. The storm-tones died away, and, turning toward the east, I beheld the countless hosts of the forests hushed and tranquil, towering above one another on the slopes of the hills like a devout audience. The setting sun filled them with amber light, and seemed to say, while they listened, "My peace I give to you."[75]

Muir gains this benediction, which he shares with the trees of the forest, after electing to experience hurricane winds while swaying at the top of a 100 foot Douglas Spruce, a position he assumed in order to better hear the Aeolian music that the wild wind was performing with the trees. A more interactive concert cannot be imagined.

The beauty and holiness of nature do not offer themselves to the mere spectator, to the theoretical gaze, to the inspector of scenery, but to the pilgrim, to the seeker, to the ascetic who leaves behind, even if temporarily, the conventions and convenience of the urban world and becomes immersed in nature. (Wendell Berry: "A man enters and leaves the world naked. And it is only naked—or nearly so—that he can enter and leave the wilderness.")[76] And if we return without the awareness that we have been addressed, we can be suspicious that we may not have made the trip at all.

The Icon/Earth Presents a Face

Icon That the icon presents a face hardly needs documentation, especially following upon the preceding discussion. However complicated the events presented in the icon, the focal point is always one or more persons—persons who address us, either directly through the orientation of their countenance, or indirectly (by proxy, as it were) through the way they address others. That the icon faces us is, in fact, a corollary and further elaboration of the third feature discussed, that the icon engages us in a reciprocity: The reciprocity is one of facing and being faced, even when the figures do not directly look out at us. For example, the right hand may be formed in a gesture of benediction to bless us, or in a gesture of silence to urge us to quiet, or they may hold an open book, asking us to recall a passage of scripture that is partially legible on its pages. In the "Hodigitria" icon, the gaze of the Virgin Mary is directed slightly away from the viewer, while her open right hand directs the viewer toward Christ who is enthroned on her cradled left arm: "She has a quiet contemplative gaze of inward attentiveness, knowledge and love, drawing the beholder into the mystery of the divine presence that is manifest in the Son she presents to the world."[77] And in the crucifixion icon, Christ looks down and to the right, even as the outstretched arms embrace the viewer and his or her world.

By facing us, the icon not only engages us, but addresses us.

Earth That we humans have always seen nature as presenting us with a face is shown by the ubiquity of what Ruskin called the pathetic fallacy: the tendency to see nature as exhibiting feelings. But it is strange to regard this as a fallacy given that it is hard to find either poetry or prose concerning the human encounter with nature that does not exhibit this tendency. (Likewise, the elimination of locutions presenting some kind of diction of nature would expunge entire libraries of poetry.) "I see something," writes Annie Dillard, "or something sees me, some enormous power brushes me with its clean wing."[78] This is apophatic phenomenology, a precise description of an encounter, not a frenzied distortion of the poet's reason by her own feelings, as Ruskin would have it.

Dufrenne tries to legitimize this experience of nature by granting it an "aesthetic" status, which seems instead merely to subvert the legitimacy of the aesthetic by rendering it non-veridical:

> As a result of aesthetic experience, something human is revealed in the real, a certain quality by which things are consubstantial with man, not because they can be known, but because they offer to the man capable of contemplating them a familiar face in which he can recognize himself without having himself composed the being of this face. Thus man can recognize his own passions in an ocean storm, his own nostalgia in an autumn sky, and his ardent purity in fire.[79]

But surely this is unsatisfactory. If the face of nature is merely our own face cast back at us, then nature is really as faceless as the back of a mirror, and nature's facing us has no more significance than the reflection of our faces by the surface of water—an optical illusion and nothing more.

Heidegger does better in this regard, maintaining at length the ontological legitimacy of the attribution of "face" and "moods," and even "saying" to nature as it is disclosed poetically, but perhaps only at the risk of reducing this capacity of nature into such general characteristics of Being as, for example, *Stimmung* and *Sagen*, while evading the really interesting and important question: If it is Being that through its own *Stimmungen* teaches us what it means to be attuned in moods, and if it is the *Logos* character of Being that first teaches us about gathering together into coherence through its own primordial saying, doesn't this suggest that Being itself somehow presents us with a face, and ultimately refers us not to an "itself" at all but to a "Thou" who demands that we pass from ontology and aesthetics to theology?[80]

The Contextuality of the Icon/Earth

Icon Every icon is situated in the midst of an ongoing, open-ended narrative that is itself composed of a plurality of narratives—a

Heilsgeschichte that embraces the many narratives of the Old and New Testaments, includes prehistories of cosmic rebellion and fall in the celestial orders, incorporates oral histories passed down from millennia, and continues with later tales of saints and martyrs that are held to be unfolding even today—and only within this contextual order is it meaningful. Moreover, narratives are seen iconically, as images of other narratives. The burning bush, and the youths in the fiery furnace, become an image or "type" of the Virgin Mary containing the uncontainable, without becoming immediately consumed, who in turn becomes the type for the faithful receiving the holy fire of the Eucharist. Jonah emerging from the belly of the whale images the risen Christ rising from the tomb. The tree of life in Eden images the tree upon which Christ was crucified and Eden restored. Indeed, this inter-textuality is part of how (as discussed in the preceding section, "The Icon Is Transactional") the faithful are inscribed within the icon: making the connections, seeing the inter-relations, the believer becomes part of the narrative.

Several examples, drawn from many more, of the use of cave imagery in iconographic themes can serve to exemplify this inter-textuality. The Nativity icon shows the infant Christ being born in a cave, indicating the earth itself taking on the glory of divinity through the Incarnation. In the Theophany icon, where Christ is baptized in the River Jordan, the river banks assume a distinctly cavelike shape to indicate a transformation of earth and a blessing of the flowing, baptismal water of the river, imaging the Spirit moving "upon the face of the waters" in the first creation, and thus signifying the restoration of Paradise. In the Crucifixion icon, a small cave under the cross (planted on Golgotha, the "place of the skull") displays a single skull—that of Adam whose bones are still merely part of the dust of the earth from which they were formed, yet whose Fall is being redeemed at that moment. And in the Resurrection icon, Christ is shown in a vast cave deep within the earth, triumphantly raising up two figures from the dark pit below: The scene is Hades, the fallen sub-terra into the midst of which Christ descends on Holy Saturday to bring light and redemption and transfiguration, and the figures are Adam and Eve in their liberation. Similar examples could be given using other elementals—water, mountains, light, and so on. The icons manifest spiritual realities in part through their contextual inter-relations—graphic, poetic, narrative, ontological, and spiritual—within a world seen variously through other icons. Phillip Sherrard, in *The Sacred in Life and Art*, makes this point well:

> An icon is not something which can be regarded as a self-contained whole, complete in itself. We have become accustomed to looking at works of art as independent entities, areas of line and colour cut off from surrounding space, enclosed in a frame and hung up on the wall. The icon is not like this. It cannot be separated off in this fashion. On the contrary, it is something whose full nature cannot be understood unless it is seen in relationship to the organic whole of the spiritual structure of which it forms a part. Divorced from this whole, hung in a frame upon a wall, and looked at as an individual aesthetic object, it is divorced from the context in which it can function as an icon. It may then be attractive piece of decoration, but as an icon it ceases to exist.[81]

Earth A complete catalogue of the modes of contextuality that are at play in our experience of nature would be considerable. Nature, after all, is boundlessly extensive, both in space and in time, and its contextuality is both synchronic and diachronic. It is not coincidental that what is now termed "ecology" was until recently called "natural history," for it is hard to spend time in the natural world without an awareness that stories are everywhere, interconnecting and interpenetrating. John Muir has also reminded us that we cannot seize hold of any single thing in nature, for everything is "hitched" to everything else, and the science of ecology has conceptualized this using the cybernetic notion of "system," an idea that is grounded in our experience of the environment, even as it leaves the latter behind in pursuit of mathematical models. Yet even in the most theoretical regions of modern natural science (for example, evolutionary theory in biology and plate tectonics in geology) narratives are still indispensable.

This overwhelming contextuality is taken as problematic by Alan Carlson, who argues that in order for the canons of "aesthetic appreciation" to apply to nature, we need more than this swirl of interrelationships, for which he employs, following Yi-Fu Tuan, the Jamesean appellation of a "blooming budding confusion." In order to procure a frame or boundary that would define a proper aesthetic object, Carlson commends scientific knowledge as providing the "foci" for constituting environmental unities that could then be appreciated.[82] But this seems wrong for several reasons. First, to make scientific concepts the basis for aesthetic experience is to point in the wrong direction, away from the embodiment entailed by *aisthēsis* and toward the abstractions of models and concepts. Despite the usage popular several decades ago, "conceptual" art remains an oxymoron. And for the same reason, art framed by concepts and theories is at best merely illustrative or didactic or hortatory. Second, it is simply counterintuitive to say that our apprehension of natural beauty is dependent upon scientific knowledge, and in fact many of our best nature writers (including Mark Twain, John Burroughs, Walt Whitman, and Lewis Thomas) have emphasized the *dangers* of scientific knowledge for the lover of beauty in nature.[83] Finally, Carlson's approach diminishes the importance of individual involvement in apprehending natural beauty (discussed in the section "The Icon Is Transactional") and opts instead for the detachment that is characteristic of the fine arts model. But the latter is unsatisfactory not only for nature, but also for most art, because throughout all but the last few centuries of human history, art has been primarily sacred art, art disclosing the holy, and not an "object" for "appreciation." Indeed, one of the main tasks of this chapter is to show the inner connections between sacred art and natural beauty that are obscured by the modern model of "art appreciation." But if scientific knowledge is inappropriate as the basis for relevant contextualities in nature, what is better suited for this role?

Annie Dillard, in her "pilgrimage" at Tinker Creek, draws upon a more poetically significant mode of contextuality by mapping natural epiphanies (for example, the "tree with the lights on it") as types and archetypes. She then deploys these in setting up a mirroring contextuality, within which events and disclosures yield additional layers of meaning. In effect, Dillard sets up a world of personal *mythoi* as part of her overall poetic project, which she then shares with her readers—most of whom find Dillard's nature fascinating, even as they realize it is not a realm they can inhabit themselves in anything but a vicarious way. Thoreau, whose *Walden* serves in many ways as a model for Dillard, attempted something similar in his earlier work, *A Week on the Concord and Merrimac Rivers*—incorporating travel narrative and poetic insight with tales drawn from Indian mythology and New England history—but failed to avoid the incumbent dangers of contrivance, arbitrariness, and ultimately incoherence. Hölderlin's poetry makes a parallel attempt, but in a more far-reaching manner, as he employs figures and tales from Greek, Hebrew, Christian, and Teutonic contexts to uncover relations and meanings in the earth and sky of his poetic world. Yet it may be said that Hölderlin, too, ends up with a private construction, one that others may find inspiring and indeed revealing, but that most (even with the aid of Heidegger's elucidations) will ultimately regard as idiosyncratic in an analogous sense to the far more elaborate assemblages of images and narratives we find evoked in Pound's *Cantos* and Joyce's *Finnegan's Wake*.

Not surprisingly, the nature writers most successful in this project are those who write from within a cultural context whose traditional narratives are still somewhat vital and intact, and for the most part this turns out to mean Native American writers. N. Scott Momaday, for example, powerfully weaves together biography, history, and descriptive prose with traditional Kiowa stories in *The Way to Rainy Mountain* in a way that succeeds where Thoreau (in *The Concord and Merrimac Rivers*) fails, simply because he writes from within a tradition of living *mythoi*. The same can be said for Leslie Marmon Silko, who has written insightfully on the power of pictographic images in language that is suggestive of iconographic depiction ("A 'lifelike' rendering of an elk would be too restrictive. Only the elk is itself. A *realistic* rendering of an elk would be only one particular elk anyway.") as well as on the importance of oral traditions for understanding "landscape" ("So long as the human consciousness remains *within* the hills, canyons, cliffs, and the plants, clouds, and sky, the term *landscape*, as it has entered the English language, is misleading.").[84] "Whatever the event or subject," she writes, "the ancient people perceived the world and themselves within that world as part of an ancient continuous story composed of innumerable bundles of other stories." It is this kind of sacred, narrative contextuality, upon which the iconographic tradition also draws, that can best disclose nature in its beauty and in its holiness, and in a way that neither scientific knowledge nor cultural inventiveness and learned allusion can even well approximate. And of course, our own Western world is the first to have abandoned these traditional, sacred narratives, so it should not be surprising that the beauty and holiness of nature elude us. It has also, of course, devastated the nature that it has thereby secularized, a theme to which the final section will return. It is, then, especially appropriate here that the one context of sacred narrative maintaining some vitality in the West today is that of Paradise and Fall, for it is the one that retains, perhaps not coincidentally, a special power for illuminating environmental considerations.

The Icon/Earth Is Paradisiacal

Icon As we have noted, iconic depiction alters the "realistic" proportions of the human body. But the forms of the surrounding environment are altered as well, and for the same reason. Just as Eastern Orthodoxy sees the Fall as having cosmic proportions, it sees the redemption and transfiguration of humanity as bound up with the transfiguration of nature as well. The trees, the animals, the surrounding landscape depicted in the icon are transformed, restored to their paradisiacal character. Their "unusual appearance," writes the great iconographer and scholar of iconography Leonid Ouspensky, alludes "to the mystery of paradise," such that everything "reflects the divine presence, is drawn—and also draws us—towards God."[85]

The cosmic implications of the Fall have always been taken seriously in the Christian East, and from early on this has been understood to entail the malign effects of human beings on the natural environment. In a remarkable text from the first decade of the eleventh century, St. Symeon the New Theologian describes an "assault of all creation," its counterattack against humanity—suspended at the last moment by divine restraint—in response to the curse that we have brought upon nature:

> When it saw him leave Paradise, all of the created world which God had brought out of non-being into existence no longer wished to be subject to the transgressor. The sun did not want to shine by day, nor the moon by night, nor the stars to be seen by him. The springs of water did not want to well up for him, nor the rivers to flow. The very air itself thought about contracting itself and not providing breath for the rebel. The wild beasts and all the animals of the earth saw him stripped of his former glory and, despising him, immediately turned savagely against him. The sky was moving as if to fall justly down on him, and the very earth would not endure bearing him upon its back.[86]

God suspends this assault, according to St. Symeon, only with the provision that humankind will serve to bring nature back to a renewed condition. In terms of the "anthropocentrism" debate in environmental philosophy, this is doubtless a kind of anthropocentrism. Human beings, although very much a part of nature here, are nevertheless a most special part. Yet this special character has nothing to do with domination, let alone consumption. Rather, humanity has the task of being the vehicle through which nature is to be redeemed. Monasteries, for example, are seen as symbolic outposts of paradise, and therefore meat is not eaten in them, just as it was not eaten before the Fall. Fasting is understood to serve, among other things, to wean us of a consumptive attitude toward nature and encourage a contemplative comportment toward "the glory of God hidden in all creation." And in the East, stories are legion of holy men and women toward whom wild animals no longer exhibit fear and antipathy. Even more radically, Olivier Clément writes that "in the sacraments, matter responds to its original vocation of being the means of communication between man and its God, and therefore among all men." It is in general, he continues, "through man, who stands between earth and heavens, that creation can fulfill its hidden sacramentality."[87] This is possible because within the earthly sphere, only human beings possess noetic understanding, for this is the way in which humans serve as divine icons. Yet this noetic vision is itself precisely the ability to see the invisible within the visible (i.e., the ability to see all creation in its iconic character). Our iconic being is manifested in our iconic seeing: We actively image the invisible divinity when we see the traces of invisible divinity in creation, and in doing so offer back that creation to its source, even while we redeem it from the Fall that we have brought about. Our unique place in nature, then, is understood not to justify the domination and exploitation of nature, but rather to allow humanity to serve as the priesthood of nature, consecrating it by our joy in its beauty.

This noetic vision, then, this iconic seeing, this seeing of the invisible within the visible, is just as much the ability to apprehend beauty: the beauty of the icon and the beauty of the earth as well. Referring to the burning bush seen by Moses, and perhaps also obliquely invoking Heraclitus and the Stoics, St. Maximos says that "the unspeakable and prodigious fire hidden in the essence of things, as in the bush, is the fire of divine love and the dazzling brilliance of his beauty inside every thing." The *logoi* of created things, the presence of the invisible within them, is at the same time their hidden beauty that can be apprehended by noetic vision. It is not by accident that the Septuagint Greek text of Genesis I uses *kalon* rather than *agathōn* to render the Hebrew, which itself contains both meanings: After each act of creation, the Creator saw that it was *beautiful*. The beautiful, then, is "a shining forth, an epiphany, of the mysterious depths of being"—the visible illuminated by the invisible. Sacraments, icons, liturgies, and the lived experience of God in nature all manifest the *kosmos noētos* through the *kosmos aisthetos*.[88] All are part of the shared redemption of humanity and nature through the disclosure of divine beauty. It is this vision, not a private predilection nor an effete aestheticism, that Dostoevsky expresses when he writes, in his sketchbook, that "beauty will save the world."[89]

Earth The narrative idea of Paradise, and of a Fall from Paradise, that would account for the present juxtaposition of humanity and nature, is not only alive today, but in one form or another it has become the leitmotiv of environmental consciousness. Within the canon of American nature writers, it is hard to find texts that do not in some way understand wilderness as a (lost) paradise, from which we have exiled ourselves.[90] Edward Abbey, to give only one example, in discussing the question of what is meant by wilderness, rejects formal definitions couched in phrases such as "minimum contiguous acres of roadless space," and chooses instead to identify it in terms of a certain love, a certain "justified. . . nostalgia for the lost America our forefathers knew":

> The word ["wilderness"] suggests the past and the unknown, the womb of earth from which we all emerged. It means something lost and something still present. . . . But the love of wilderness is more than a hunger for what is always beyond reach; it is also an expression of loyalty to the earth, the earth which bore us and sustains us, the only home we shall ever know, the only paradise we ever need—if only we had eyes to see. Original sin, the true original sin, is the blind destruction for the sake of greed of this natural paradise which lies all around us—if only we were worthy of it. . . . The Paradise of which I write and wish to praise is with us yet, the here and now, the actual, tangible, dogmatically real earth on which we stand.[91]

It is important here that although Abbey concludes with a strong emphasis on paradisiacal nature as concrete and actual rather than ideal ("the here and now, the actual, tangible, dogmatically real earth on which we stand"), at the same time he notes (as does the Genesis narrative) two barriers barring our way back into Paradise, and these constitute our "true original sin": an inability to see ("if only we had eyes to see") and our "blind destruction for the sake of greed of this natural paradise which lies all around us—if only we were worthy of it." To regain the paradise "which lies all around us," we must forsake our avarice for worldly gain, undergo a kind of *askēsis*. But at the same time, we must overcome our inability to see this sustaining, surrounding paradise, and it can even be inferred that since the destruction is "blind destruction," regaining our ability to see is more primary. Several pages later, Abbey quotes from the "panegyric account" of John Wesley Powell on the "glories and beauties of form, color, and sound" in the Grand Canyon, that offer up a "sublimity" not to be found "on the hither side of Paradise." Yet Powell cautions that "you cannot see the Grand Canyon in one view, as if it were a changeless spectacle from which a curtain might be lifted, but to see it you have to toil from month to month through its labyrinths."[92] The vision of paradise, then, requires effort and sacrifice.

And what is this vision of paradise? It is simply the beauty of nature, of the invisible shining forth and illuminating the visible. But of course this is not a simple property or attribute, a "spectacle" to be glimpsed "in one view" as if "a curtain might be lifted." It is instead that toward which all the writers discussed here have been pilgrims, toward which their searchings and strivings have been oriented. It explains why, in diverse ways, each has had to become a pilgrim and ascetic. It explains why the beauty and the godliness of nature cannot be adequately understood apart from one another, and why the most ardent seekers of the beautiful in nature have been at the same time, in one way or another, spiritual seekers as well. It explains why the style of their writings, parallel to iconic styles, is epiphanic and apophatic, rather than mimetic or anamnetic. And it connects Thoreau's vision of Paradise, through which "wildness"—both in nature and in the soul—is looked to for "the preservation of the world," to that of his Russian contemporary Dostoevsky, cited already and also much concerned with Paradise, who looks toward "beauty" to "save the world."

The Iconic Earth in the Work of Dostoevsky

> In adolescence, when Dostoevsky's characters first began to impress me with the violence of their tragic mysticism, I knelt

> before the icon of the Virgin that sat enthroned above my bed and attempted to gain access to a faith that my secular education did not so much combat as treat ironically or simply ignore. I tried to imagine myself in that enigmatic other world, full of gentle suffering and mysterious grace, revealed to me by Byzantine iconography.
>
> *Julia Kristeva,* In the Beginning Was Love: Psychoanalysis and Faith[93]

Within a twenty-year period, from the early 1870s to the late 1880s, both Nietzsche and Dostoevsky were preoccupied with the question of European nihilism. Both regarded nihilism as involving a misrelation to the earth, both saw the question of divine transcendence as a key to understanding it, and both saw the overcoming of nihilism as having an aesthetic dimension. Their diagnoses, however, are diametrically opposed, and it is not surprising that the analysis of Nietzsche, the self-avowed "good European" whose thought is deeply rooted in modern European concerns, should have been embraced one-sidedly in Western intellectual circles, to the near exclusion of Dostoevsky's diagnosis, which looks at Western nihilism from the East. But if this scholarly imbalance is not surprising, neither is it felicitous, especially with regard to understanding the relationship of environmental crisis to contemporary nihilism. Both diagnoses, it could be argued, proceed from the Western over-emphasis on divine transcendence that has been discussed previously. But whereas Nietzsche's Zarathustra exhorts us to be true to the earth by rejecting transcendence altogether, Dostoevsky's most prophetic characters in his later novels commend a vision of the earth as once again infused with the transcendent, a vision of the earth as sacramental and iconic. Redemption for Nietzsche can be seen as the work of human creators willing to assume the place of God. Whereas for Dostoevsky it calls for the rejection of this same modern project of attempting to create self and world (and whose disastrous, abortive effects are studied in *Devils*), and calls instead for seeking, in repentance and humility, the image of divine beauty in humanity and in nature, as exemplified in *The Brothers Karamozov* by Alyosha and by the Elder Zosima—Dostoevsky's rendering of the actual St. Tikhon of Zadonsk.

In *Devils*, previously translated as *The Possessed*, and whose final installment was published in 1873, nihilism is presented as consisting of interrelated negations: the denial of transcendence, contempt for the life of the peasants, and a disdain for the earth.[94] (Shatov's injunction to the arch-nihilist and smug child-molester Stavrogin: "'Kiss the earth, flood it with tears, ask forgiveness! . . . acquire God by labor; the whole essence is there, or else you'll disappear like vile mildew; do it by labor . . . peasant labor. Go, leave your wealth.'")[95] In *A Writer's Diary*, the July and August entries of 1876 include a section called "The Land and Children." Dostoevsky anticipates Aldo Leopold and Wendell Berry here in arguing that we need to live closer to the land, maintaining that the earth is "sacramental," introducing the theme of the earth as Paradise or "Garden," and insisting that children "should be born and *arise* on the land, on the native soil in which its grain and its trees grow."[96] And in the April entries of the following year, he publishes serially his remarkable short story, "The Dream of a Ridiculous Man," termed by Mikhail Bakhtin an "encyclopedia of Dostoevsky's most important themes."[97] The story commends to the reader a "living image" of the earth as paradise—of an earth "which seemed to have a festive glow, as if some glorious and holy triumph had at last been achieved" (i.e., a vision of what we have called the iconic earth). Not only do its inhabitants love one another, they also love the earth, singing hymns of "nature, the earth, the sea, and the forests," and conversing with the very trees with an "intensity of love." "They regarded the whole of nature in the same way [as the trees]—the animals, which lived peaceably with them and did not attack them, conquered by their love" and the stars as well. The dreamer relates how in homage to this unfallen world, he "kissed the earth on which they lived." On awakening, however, he realizes that this "living image" of paradisiacal earth is not just a dream but "the truth," and that it could be realized "at once" if we truly wanted it.[98]

In *The Brothers Karamazov*, completed three years later, Dostoevsky felt that he had finally arrived at a statement that pointed beyond the nihilism which threatened not only Western Europe but the entire earth. Here he develops the themes of the preceding works extensively and powerfully, elaborating them explicitly within the sphere of Orthodox spirituality, and thus only a very brief and general characterization of them is possible here.[99] Presenting his final teaching, the Elder Zosima recalls how his adolescent brother had undergone a wondrous change as he neared an untimely death. Remorseful over his previous bitterness and cynicism, yet weeping with joy, he asks for forgiveness even from the birds singing outside his window: "Birds of God, joyful birds, you, too, must forgive me, because I have also sinned before you . . . Yes . . . there was so much of God's glory around me: birds, trees, meadows, sky, and I alone lived in shame, I alone dishonored everything, and did not notice the beauty and glory of it all." Awakened, his eyes opened, he now sees that "life is paradise, but we do not want to know it, and if we did want to know it, tomorrow there would be paradise the world over."[100] Thus, a kind of nihilistic unbelief prevented him from seeing the paradisiacal glories of the nature around him, and only when he repented of this were his eyes open. Zosima then relates how he himself, repenting of pride that had brought him to the brink of a duel, had cried out with a new awareness: "Look at the divine gifts around us: the clear sky, the fresh air, the tender grass, the birds, nature is beautiful and sinless, and we alone, are godless and foolish and do not understand that life is paradise, for we need only to wish to understand, and it will come at once in all its beauty."[101]

The Elder goes on to deliver a series of "talks and homilies" to his closest friends and students. In these, he presents his understanding of the dangers of Western nihilism and of the iconic vision that can overcome it. He warns against scientific materialism, which distorts the image of God in nature; against the moral relativism that he sees engendered by science; against the acquisitive materialism, individualistic hedonism, and "spiritual suicide" of the newly wealthy; and against the age's rejection both of "the spiritual world" and of "the idea of serving mankind, of the brotherhood and oneness of people." Exhorting repentance of this nihilistic mindset, he urges instead love not only for other people but for nature as well, in an admonition that could well serve as the quintessential environmentalist sermon, and that explicitly evokes the epigram from St. Isaac of Syria—whose *Ascetical Homilies* became Dostoevsky's companion—that stands at the beginning of the chapter:

> Brothers, do not be afraid of men's sins, love man also in his sin, for this likeness of God's love is the height of love on earth. Love all of God's creation, both the whole of it and every grain of sand. Love every leaf, every ray of God's light. Love animals, love plants, love each thing. If you love each thing, you will perceive the mystery of God in things. Once you have perceived it, you will begin tirelessly to perceive it more and more of it every day. And you will come at last to love the whole

world with an entire, universal love. Love the animals: God gave them the rudiments of thought and an untroubled joy. Do not trouble it, do not torment them, do not take their joy from them, do not go against God's purpose. Man, do not exalt yourself above the animals: they are sinless, and you, you with your grandeur, fester the earth by your appearance on it, and leave your festering trace behind you . . . My young brother asked forgiveness of the birds: it seems senseless, yet it is right, for all is like an ocean, all flows and connects; touch it in one place and it echoes at the other end of the world. Let it be madness to ask forgiveness of the birds, still it would be easier for the birds, and for a child, and for any animal near you, if you yourself were more gracious than you are now, if only by a drop, still it would be easier. All is like an ocean, I say to you. Tormented by universal love, you, too, would then start praying to the birds, as if in a sort of ecstasy, and entreat them to forgive you your sin. Cherish this ecstasy, however senseless it may seem to people. My friends, ask gladness from God. Be glad as children, as birds in the sky.[102]

Nature Godly and Beautiful

And I stamped my foot angrily: "Are you not ashamed, unhappy animal, to whine about your fate? Are you not able to free yourself of subjectivity? Are you not able to forget yourself?" . . . Objectivity does exist. It is God's creation. To live and feel together with all creation, not with the creation that man has corrupted but with the creation that came out of the hands of its Creator; to see in this creation another, higher nature; through the crust of sin, to feel the core of God's creation. . . . But to say this is to posit the requirements of a restored, i.e., a spiritual person. Once again, the question of asceticism arises.

Pavel Florensky, The Pillar and Ground of the Truth[103]

The soul cannot get enough of beholding the beauty of nature.

Elder Ephraim[104]

What are the theological and aesthetic roots of environmental crisis? Following Nietzsche, we could say that they lie in the overemphasis on transcendence that he believes was inaugurated by Platonism, and then appropriated by Christianity. But twelve decades later, as belief in divine transcendence has progressively dimmed, environmental problems have gotten steadily worse, and rather than gaining joy in the earthly, we have embraced pleasure in consumption. Moreover, this diagnosis fails to explain why this crisis has been associated with Western Christianity, rather than Judaism or Islam, both of which have traditionally placed even greater stress on divine transcendence. Alternatively, we could follow Heidegger and look toward that vanishing point where a bias toward "presence" first emerges at the beginning of Greek philosophy, but this etiology alone fails to explain why it took almost two thousand years for the symptoms to appear, and in addition threatens to render Western civilization—now become global—so thoroughly questionable that it leaves us waiting in silence for nothing less than a last God and a new beginning. Alternatively, the roots of environmental crisis have been located in patrilineal cultures worshipping male gods (Gary Snyder, and some eco-feminists), and in the rise of phonetic writing (David Abram). But these analyses are subject to the same objections as is Heidegger's: Setting aside practical questions concerning their implementation, both fail to explain why a distinctive set of beliefs, attitudes, and practices toward the environment began to emerge several hundred years ago, rather than several thousand years ago.

Lacking compelling reasons to the contrary, it is better to look for roots where the problem first begins to become manifest, and this is the period on which Lynn White Jr. focuses, the Latin Middle Ages. During this time, the onto-theology that is the proximate subject of Heidegger's critique took its definitive shape, and the unmitigated transcendence to which Nietzsche objects begins to be valorized, a process eventually leaving the West with a deistically vacated *oikos*, or household, that could be appropriated without restraint by its new, human landlords: a secular nature devoid of divine immanence. At the same time, as art became secularized as well, the last visible window to the divine was closed, and spiritual experience was confined to the spheres of human subjectivity (and eventually, emotion alone) and onto-theological metaphysics.

Is this the "destiny of the West," or is it rather a defection from its own guiding insights, a deformation of what is most basic and characteristic? If *the West* means something like that distinctive synthesis of Roman law, Greek philosophy, and Middle Eastern spirituality that took shape in the ancient world, then "the East" of Byzantine art and spirituality can surely lay a more legitimate claim to being truly *Western* than can "the West." For Byzantium, New Rome, preserved and developed that synthesis for a thousand years after the collapse of Rome in the fifth century—a disintegration persisting for almost four hundred years until the "Carolingian Renaissance" inaugurated what largely became a new direction for Western Europe.

Regarding "the West" from this perspective, as undergoing a kind of default from the truth of its own beginnings, the view is altered dramatically. We see the steady unfolding of a secularized, disenchanted, and ultimately mechanized natural environment along with the art and artifice that are increasingly devoted to its mastery and exploitation, a development that is nevertheless punctuated by the dissent of a few odd religious figures (St. Francis of Assisi, Meister Eckhart, Jacob Boehme); a series of strange poets (Blake, Hölderlin, Wordsworth); and a lineage of eccentric naturalists (Thoreau, Muir, Abbey, Dillard)—all struggling to retrieve the noetic vision, within the earthly and visible, that was becoming obscured in the West. That is, the "Eastern" iconic orientation toward the earthly would be that self to which the West is called to be true by these visionaries, who would in this case be not dissenters at all, but rather the genuine loyalists.[105]

Working independently, both Philip Sherrard and Seyyed Hossein Nasr have blamed our present environmental crisis on our Western establishment of the first truly secular culture the world has known. Nasr calls for a "resacralization of nature, not in the sense of bestowing sacredness upon nature, which is beyond the power of man, but of lifting aside the veils of ignorance and pride that have hidden the sacredness of nature from the view of a whole segment of humanity."[106] And Sherrard calls for a recovery of "our capacity to perceive this symbolic function of natural things—to perceive the numinous presence of which each natural form is the icon—that [was] increasingly eclipsed by those intellectual developments that took place in the Christian, and hence by and large European, consciousness in the later medieval period."[107] For Sherrard and Nasr, the modern West is anomalous and apostate from the otherwise shared experience and wisdom of humankind according to which, in the words of Elder Zosima, this world "lives and grows only through its being in touch with other mysterious worlds."[108] If this is in fact the case, it should not be surprising that principal structures of the iconic vision sustained by Eastern Orthodoxy would be found at play in Western experience. In view of the iconic character of Latin art and spirituality prior to the Middle Ages discussed previously, this iconic sensibility would represent the last point of contact for Western culture with humanity's common experience of nature as godly and beautiful.

There are practical and prescriptive issues here, too, alongside the scholarly and diagnostic questions. Environmental thought needs to take more seriously both the aesthetic and the theological dimensions of nature, and it cannot understand the former without coming to terms with the latter. Only through its willingness to take up aesthetico-theological questions can environmental philosophy proceed beyond the limitations of the ethical. And until the beauty of nature is acknowledged in its numinous character, the natural environment will not be granted the respect, and indeed veneration, it deserves as a primary locus of what John Muir called "endless, inspiring, *Godful* beauty."[109]

Beyond this, it seems unlikely that the mindless consumption and exploitation that have so far devastated the earth, and whose acceleration has become global, will be genuinely reversed unless the turn is grounded in a mode of spirituality that can successfully balance transcendence and immanence, while maintaining the vigor to make serious practical demands upon us. Snyder, aware that the beliefs and practices of "primary peoples" will not be widely practicable, feels that Buddhism can best accomplish this task, and some in the West have found this to be a viable way.[110] Nasr takes the Sufi position that each of the world's great religious traditions—if pursued with sincerity and in their traditional forms—will ultimately (but not initially) converge upon the same esoteric vision, including a sacred knowledge of nature.[111] It does not, however, seem likely that the sensibilities of New Age spirituality will accomplish this.

Finally, the understanding of the earth as iconic helps us to articulate perhaps the most important basis of all for justifying environmental preservation, one that is intuitively evident already to most environmentalists. Of the icon, Massimo Cacciari has written:

> The passage from the Invisible to the Visible and from the Visible to the Invisible, forms the icon's liturgical and sacramental essence; it is the royal gate through which the Invisible is manifested and the Visible is transfigured . . . Wanting the destruction of an icon is like "walling a window" or like obstructing the passage though which Light overflows into our world and comes before our sensitive eyes.[112]

Likewise, we must be able to argue that the destruction of the earth's wild species and pristine places is odious not just for its depriving certain people of recreational opportunities to which they happen to be devoted, nor is it even merely for its decreasing of the aggregate of beauty on the earth, nor for depriving scientists of species to study and pharmacists of pharmaceutical materials. It is, rather, the progressive walling of windows, windows that we did not build and cannot replicate, and that will thus be walled-over forever.

CHAPTER 7

Seeing Nature

Theōria Physikē in the Thought of St. Maximos the Confessor

> Much on earth is concealed from us, but in place of it we have been granted a secret, mysterious sense of our living bond with the other world, with the higher heavenly world, and the roots of our thoughts and feelings are not here but in other worlds. That is why philosophers say it is impossible on earth to conceive the essence of things. God took seeds from other worlds and sowed them on this earth, and raised up his garden; and everything that could sprout sprouted, but it lives and grows only through its sense of being in touch with other mysterious worlds; if this sense is weakened or destroyed in you, that which has grown up in you dies. Then you become indifferent to life, and even come to hate it. So I think.
>
> —*Elder Zosima in Dostoevsky's* The Brothers Karamozov[1]

> So the soul flees toward the intellectual contemplation of nature, as to the inside of a Church and to a place of peaceful sanctuary . . . And there it learns to recognize the essential meanings of things as if through the readings from Holy Scripture.
>
> —*Maximos the Confessor,* Mystagogy[2]

I

It was Heidegger who first made "the earth" a possible topic for serious philosophical inquiry. His former student, Hans-Georg Gadamer, vividly describes the philosophical "sensation" that was generated by the "new and startling" concept of "earth," as it was introduced in several 1936 presentations of what was to later become "The Origin of the Work of Art."[3] Yet Heidegger was by no means the first philosopher to use the word in a philosophical context. For example, in the work of Nietzsche, *die Erde* and *das Irdische* (earth and the earthly) play a crucial part in the unfolding of his metaphysics, even if they were never to quite become technical terms in the lexicon of Nietzsche scholarship. Thus, throughout his writings, Nietzsche adds the phrase "on earth" to many of his philosophical claims, in a way suggesting a rebuttal of the discursive analogy and metaphysical bond expressed in the great Christian Prayer: "on earth as it is in heaven." For Nietzsche always leaves out the last clause, "as it is in heaven," intending to remind us by this omission that we are "here on earth," to bring us back "down to earth," to wake us from our heavenly dreams and return us to earthly reality. We all know about Zarathustra's call to be "true to the earth" and his charge that "to sin" not against heaven, but "against the earth is the most dreadful thing."[4] We remember, too, and more darkly, the madman's frantic alarm in *The Gay Science* that the earth has been "unchained from its sun," and we may wonder in passing how it can remain earth without its sun overhead, an earth without its heavens. But Zarathustra answers this doubt with a surprising claim and a peculiar imperative: "The overman is the meaning of the earth. Let your will say: the overman *shall be* the meaning of the earth!"[5] This is not what environmental philosophers would most like to hear. "The overman *shall be* the meaning of the earth!" This is not deep ecology. We could call it anthropocentrism were it not that the overman is supposed to be more than *anthrōpos*: more creative and more masterful. Colder and harder. More domineering. More dominant.

Environmental thinkers have tended to uncritically embrace Nietzsche's claim that metaphysics (especially Platonism) and theism (especially Christianity) have, in the language of *Beyond Good and Evil*, "cast suspicion on the joy in beauty." But they have also read Nietzsche rather selectively, in this case ignoring the remainder of the same sentence: "bend everything haughty, manly, conquering, domineering, all the instincts of the highest and best-turned-out type of man, into unsureness, agony of conscience, self-destruction—indeed invert all love of the earthly and dominion over the earth into hatred of the earth and the earthly."[6] Haughtiness? Testosterone? Conquest? "Love of the earthly *and domination over the earth*"? How to philosophize with a bulldozer? Or so Heidegger himself at last came to read Nietzsche. In the final volume of his book on Nietzsche, he argues that the rejection of the heavenly and invisible and transcendent—of "the old 'traditional' valuation [that gave] to life the perspective of something "suprasensuous, supraterrestrial—*epekeina*, 'beyond'"—this rejection leads in Nietzsche not to the virtues that environmentalists hold dear, but rather to "unrestricted" and "all-encompassing" and "absolute domination of the earth."[7] Indeed, it is in just this way that the overman redeems the earth, *becomes* its meaning, by imposing his own aims upon the earth: Divine dominion is exchanged for total human domination. By walling-off the heavenly and restricting the earth to one-dimensionality, the power of human will becomes focused and intensified and totalized. Denied its transcendent outlet, the human will-to-power waxes and builds here "on earth." The resultant humanity will have, Heidegger's exegesis continues, an "essential aptitude for establishing absolute dominion over the earth. For only through such dominion will the absolute essence of pure will come to appear before itself, that is to say, come to power."[8]

Furthermore, if Heidegger is right about Nietzsche's thought of an earth disconnected from its heaven, then a key presupposition of recent environmental thought becomes dubious. Lynne White Jr., in an essay discussed already in the third and sixth chapters of this book, and that I have argued has had a widespread and ultimately toxic effect upon environmental thought, concludes that it is the sense of human transcendence (based upon the human imaging of a transcendent God) that has made us feel entitled to dominate the earth. This thesis, rarely questioned in environmentalist circles, has been used to justify various virulent strains of biological and ecological reductionism, and it has made antipathy to theistic spirituality *de rigueur* among right-thinking environmentalists. But if Heidegger is right, then Lynn White has gotten it backwards: Precisely the *lack* of transcendence has established the imperative to dominate as an ontological cornerstone of our contemporary world. We return here to Marcel Gauchet: "As God withdrew, the world changed from something *presented* to something *constituted*. God having become Other to the world, the world now becomes Other to humans. . . . Disentangling the visible from the invisible made it 'inhuman' in our minds, by reducing it to mere matter. At the same time, this made it appear capable of being wholly adapted to humans, malleable in every aspect and open to unlimited appropriation."[9]

II

Gauchet is not characterizing here the program of Nietzschean metaphysics alone, but rather the much broader dynamic that

Max Weber, under the influence of both Hegel and Nietzsche, called the *Entzauberung*—the dis-enchantment, or better, de-magification—of the world. When the gods withdraw, or are driven out, from nature, then the world becomes disenchanted, demystified, secularized. We can do with it as we please, have our way with the earth and the earthly. And it is largely Christianity that is claimed to have performed this cosmic exorcism.

But there are strong reasons for seeing this as a half-truth at best. *First*, if we refrain from romanticizing the divinities of the ancient Greeks and Romans, we can readily see that their inherence in nature was regarded as more menacing than inspiring: All the various duties of *religio*—sacrifices and libations and all the rest—were performed not out of love or awe or even respect, but more as placation and propitiation, as concessions to cosmic bullying. Even Plutarch, himself an ordained priest of Apollo, conceded that "these feasts and sacrifices were instituted only with the aim of sating and appeasing the evil demons," i.e., the personages and powers of the Greek pantheon.[10] Prior to the advent of the mystery religions, there is scant evidence that anyone had taken these practices very seriously for some time, apart from assuring that they were performed correctly. Thus, there is little reason to think that nature exhibited any real "enchantment" during the classical periods of pagan antiquity, either regarding the gods per se, or their inherence in nature: As is made abundantly clear in both the *Iliad* and the *Odyssey*, as well as Hesiod's *Works and Days*, the *kosmos* was more like a treacherous minefield than it was like a wondrous, enchanted forest. And so it follows that the ancient Greeks and Romans had no "nature poetry" in the modern sense.

Second, as Peter Brown has shown in his studies of late antiquity, if Christianity dispelled the pagan magic, it more than compensated by substituting its own sense of the sacred in nature, one that was even more powerful—and certainly more heartfelt—than its predecessors. The sacred places in nature now become the sites of historical epiphanies, places where holy men and women enabled the connecting circuit between the visible and the invisible to be powerfully manifest: places of serene martyrdom and miraculous deeds and theophanous events that rendered the surrounding places and landscapes themselves to be enduringly charged with the divine energies.[11]

But there is a *third* set of reasons that the "disenchantment" thesis obscures and underestimates the positive contributions of Christian philosophy and spirituality, and it is the consideration of these that will form the central concern of this chapter. I will argue that rather than substitute an abstract, rationalized, and exterior set of relations between humanity and nature for the *daimonic* interiority that it had exorcised—a process that came much later with scholastic philosophy and the subsequent rise of modern science and technology—the deepest currents of traditional Christian thought and spirituality during the first millennium instead *strengthened* and *more deeply interiorized* this element, so much so that it eventually served to make possible the most vital and most important sensibilities of modern environmental thought. This is, of course, a far-reaching and seemingly counterintuitive thought, and the remainder of this chapter is devoted to a preliminary effort to make it intelligible—and perhaps even somewhat plausible as well.

III

Pavel Florensky was surely the brightest star of the cultural and intellectual renaissance that took place in Russia just prior to the revolutions of 1917. Florensky was a philosopher and theologian, but also a linguist who had mastered a dozen languages, a scientist and mathematician of international renown, and an art historian and intellectual historian whose critical works are now being translated. He also had a deep and even mystical love for the natural environment, drawn partly from an idyllic childhood in the Caucasus Mountains, which remained with him during the harshest years of confinement in the Siberian *gulags*. A major chapter of his masterpiece, *The Pillar and Ground of Truth*, is devoted to nature understood as "creation," and here he develops a remarkable thesis, running directly counter to the thesis advanced by Nietzsche and White concerning nature and Christianity. Florensky argues:

> Only Christianity has given birth to an unprecedented being-in-love with creation. Only Christianity has wounded the heart with the wound of loving pity for all being. If we take the "sense of nature" to mean . . . more than an external, subjectively aesthetic admiration of "the beauties of nature," this sense is then wholly Christian and utterly inconceivable outside of Christianity, for it presupposes the reality of creation. [This] sense of nature . . . became conceivable only when people saw in creation not merely a demonic shell, not some emanation of Divinity, not some illusory appearance of God,

like a rainbow in a spray of water, but an independent, autonomous, and responsible creation of God, beloved of God and capable of responding to His love.[12]

One aspect of Florensky's thesis, as he presents it, will resonate with current scholarship, for his argument that modern natural science presupposes the belief in a *logos* operating consistently and intelligibly within the *kosmos* has by now been accepted by many historians of science. And although this thought begins with Heraclitus, and is developed by the Stoics, it took a Christian cosmological vision to arrive at the notion of a "book of nature"—a phrase first used by St. Anthony the Great in the fourth century—whose characters could be intelligibly "read." But other aspects of Florensky's claim seem at first to be far less plausible, given our usual views of the Christian tradition. "The wound of loving pity for all being"? "An unprecedented love for creation"? In the Christianity that developed in the West, especially after the Great Schism of 1054, only St. Francis comes readily to mind here. But things are different if we turn to the Christian East. Here we find St. Isaac the Syrian, a seventh century monk much admired by Dostoevsky, who expresses this cosmic love as follows:

> An elder was once asked, "What is a compassionate heart?" He replied:
>
> "It is a heart on fire for the whole of creation, for humanity, for the birds, for the animals, for demons and for all that exists. At the recollection and at the sight of them such a person's eyes overflow with tears owing to the vehemence of the compassion which grips his heart; as a result of his deep mercy his heart shrinks and cannot bear to hear or look on any injury or the slightest suffering of anything in creation.
>
> "This is why he constantly offers up prayers full of tears, even for the irrational animals and for the enemies of truth, even for those who harm him, so that they may be protected and find mercy.
>
> "He even prays for the reptiles as a result of the great compassion which is poured out beyond measure—after the likeness of God—in his heart."[13]

Mature monastic *askēsis* here arrives not at a disavowal of nature but at a cosmic embrace. And if we examine the actual development of Christian monasticism, and of the philosophical reflections it shaped, in contrast to the fictitious asceticism of Nietzsche's imagination, it is quite clear that a movement *away* from the earth-despising attitude of Platonism was not just a discernable feature, but precisely its most salient characteristic. And more generally as well, Plato's demotion of the realm of becoming was the principal reason that his writings were held at arm's length throughout Byzantine intellectual history, for the divine exclamation at the end of each day of creation that the visible, palpable, audible world was, in the words of the Septuagint, "*kalon*," i.e. "fine" in the senses of both good and beautiful.

For Plato, the visible world is remote from true being. Epistemologically, he maintains, our experience of the earthly offers only faint and distorted evidence of the real, and this is because metaphysically it is several steps removed from the real, which lies beyond it, and which can be reached only to the extent that the wings of the soul lift us above the earthly. And only at this point, beyond the earthly and visible, can we speak of a contemplation or *theōria* of true being. Perhaps the closest Plato comes to prescribing a contemplative relation to the *kosmos* is in the *Timaeus*, where it is argued that the *daimonic* part of the soul needs to be appropriately fed, and that by reflecting on the motions of the heavens, this highest part of the soul can nourish and strengthen the inherent buoyancy "that lifts us up toward our kindred in heaven and away from the earth" (Tim. 90a–d).[14] Thus, only the very peak of the visible (the light of the heavens) is of higher interest, and even then, only in order to elevate us beyond the visible into what the Phaedrus calls "that place beyond the heavens. [where] true being dwells" (Phaedrus, 247c).

"Now to discover the poet and father of the all," says Timaeus, "is quite a task" (Tim. 28c). And though the *Timaeus* goes on to argue, on the basis of cosmic beauty, that there must be a demiurge, this is presented as an inference rather than an attending, as a path to the divine that is mediate and discursive rather than in any sense contemplative or noetic. But if we look not toward Athens but toward Jerusalem, we will find something very different:

> The heavens declare the glory of God;
> and the firmament proclaims the work of His hands . . .
> Day upon day pours forth speech,
> and night upon night proclaims knowledge.

There are no tongues nor words
where their voices are not heard.
Their sound has gone out into all the earth,
and their words to the ends of the world.
(Psalm 18/19)

Whereas for Plato it is difficult, "quite a task," to discern the divine even at the very pinnacle of the natural order, for the psalmist heaven and earth speak always and everywhere of holy beauty, the divine glory. So loquacious and eloquent are these voices that in the "Wisdom of Solomon," those who nevertheless manage to overlook the divine beauty in the cosmos are depicted as not only "ignorant," but "foolish." For the "incorruptible Spirit [of God] is in all things," and thus "by the greatness and beauty of creation, proportionably the maker of them is seen [*theōreitai*]" (Wisdom 12:1–13:5). Not surmised or deduced or concluded, but *seen*. Cosmic aesthetics unfolding into a true "natural theology" that is immediate and experiential rather than discursive and inferential as it was in Greek philosophy and even more its scholastic successors.

This is a very different sensibility, a sense of nature as a field or arena that is shot through with divine energies and holy syllables, of nature not as mute and unreal and ultimately dispensable, but as solid and expressive and eminently worthy of our attention. This new sense is articulated philosophically in Middle Platonism by the Jewish philosopher Philo of Alexandria, who was the first to bring together philosophically Athens and Jerusalem, and who unlike Plato felt that it was not difficult but *easy* to apprehend the agency, and thus the existence of God in the splendor of nature, and above all in the order and beauty of heavenly nature.[15] Easy and natural, because with Philo, we find a God who is expressive, who speaks and declares Himself.[16] Yet with Philo, as with Plato, even the heavens are of only transitory interest in the ascent of the soul toward the intelligible. It is, rather, with the advent of Christianity, not just as the latest of the "mystery religions," but as a school of ancient philosophy, that nature begins to receive a philosophical legitimacy of its own, becoming worthy of contemplation in its own right.

IV

The contemplation of nature, *theōria physikē*, has a rich development that begins in late antiquity, and its history remains to be written. This "natural contemplation" is by no means the kind of discursive, explanatory inquiry into nature with which we are familiar in earlier Greek philosophy from Parmenides's account of the deceptive, second-rate world of appearances to the discursive accounts of Aristotle and the Greek Atomists. Rather, the same kind of noetic, contemplative comportment that Plato had reserved solely for the eternal forms, denuded of any earthly encumbrance, is now directed toward the cosmos itself. It is the practice of seeing (*theōrein*) the divine depths (*logoi*) of nature: of seeing nature in God, and seeing God in nature. *Theōria physikē* was developed almost entirely within the frame of reference of Christian philosophy, and it had few, if any, counterparts among the pagan philosophers—confounding to those today who believe that it was paganism that venerated nature and Christianity that turned its back upon it. Only a few figures in this development can be considered here, all but one of them in a rather cursory manner. All were mystics and philosophers of the Greek East—from Alexandria and Constantinople to the Cappadocian Highlands and the Syrian and Egyptian Deserts—and contrary to what we would expect if Nietzsche was correct about asceticism as world-negating, all five were Christian monastics.

It was the third-century Alexandrian philosopher Origen—fellow student with Plotinus at the school of Ammonios Saccas—who first employed the term *physikē* to refer not to the investigation, but to the contemplation of nature. And it was Origen who first articulated what it was that this contemplation sought—the *logoi* that inhered within all creation. For Origen took his bearings, as would the other four, from the "Prologue" to St. John, according to which it was through the eternal Logos or Divine Word that all things were created. Thus, Origen can characterize this *physikē* by stating that "He who made all things in wisdom so created all the species of visible things upon the earth that He placed in them some teaching and knowledge of things invisible and heavenly whereby the human mind might mount to spiritual understanding and seek the grounds of things in heaven."[17] Ultimately, however, Origen's pagan roots pull him back into world-dismissal, for he retains the Platonist belief that our very life within the visible is a kind of lapse, and thus concludes that the contemplation of nature ultimately serves only to prepare us for our homecoming in the invisible.

Gregory of Nyssa, the fourth century Cappadocian, further develops this contemplative approach to nature. His allegorical reading of the life of Moses explores how the contemplative life can be understood as articulating the pathway to becoming a "friend of God," and glossing the Psalm cited previously, he sees nature as an important source of our knowledge of God: "For the wonderful harmony of the heavens proclaims the wisdom which shines forth in the creation and sets forth the great glory of God through the things which are seen, in keeping with the statement, *the heavens declare the glory of God*."[18] Unlike Origen, Gregory rejects the Platonic notion of a *chōrismos* between the visible and the invisible, and he regards the inner *logoi* of creation not just as images of the divine but as divine words spoken though, and indeed *to*, each created thing—as in each case "an interior word that was spoken to these very same beings, a living voice" that is immanent in existent beings. Even the senses—so despised by Platonists of the strict observance—become, in Gregory's words, "signposts for penetrating deeper into the invisible from the visible." At the same time, Gregory is aware of the divine mystery that inheres within the deepest recesses of creation, and thus cautions that our quest for knowledge must disavow its tendency to grasp after things like a raptor seizing its prey: because of their very interiority, we can never fully capture these *logoi* that lend life to creation.[19]

Both Origen and Gregory had emphasized that *theōria physikē* required important prerequisites in life, a successful acquisition of virtue (*ethikē*) and a purification of the soul (*katharsis*) from passions such as gluttony, avarice, anger, and vainglory. But with Evagrios of Pontos, friend and younger contemporary of Gregory, we meet not only an ascetic, but a monastic who entirely abandoned the sophisticated cities of the late Roman empire for the harsh wilderness of the desert. Evagrios emphasizes more clearly than either Origen or Gregory that the mystical quest not only may, but *must* pass through the stage of natural contemplation. Nor is there anything in his writings to indicate that it need be left behind at all, even if it is to be regarded as penultimate to something even higher. He contrasts *theōria physikē* first with the ordinary *human* understanding of things, which approaches them naively and superficially, just as they present themselves at first glance, and second with a *demonic* understanding of things, which deals with things under the influence of the passions, solely in their materiality, and only in order to appropriate and exploit them. Natural contemplation, in contrast, is neither human nor demonic, but *angelic*. Like Cleopas, whose heart burned within him as the Incarnate Word revealed the inner meaning of scripture on the Road to Emmaus, Evagrios commends to us as well that our hearts burn within as we practice this angelic contemplation of the *logoi*, the inner meaning of the things themselves.[20] Evagrios holds additional importance here, because he placed particular emphasis upon the ascetic struggle to overcome the passions and attain virtue, thereby laying the foundation for all later writings on ascetism. But far from being world-denying, this effort for Evagrios has as its aim just the opposite: "We practice the virtues *in order* to achieve contemplation of the inner meaning (*logoi*) of existent things, and from this we pass to contemplation of the divine Logos *in the ontological heart of all things.*"[21]

Finally, at least brief mention must be made of Dionysios the Areopagite, nominally the companion of St. Paul and first bishop of Athens, but most likely a Byzantine monk of the late fifth or early sixth century. Although Dionysios has become celebrated for his negative or "apophatic" theology in his *Mystical Theology*, he also developed a positive or kataphatic theology as well, intended as penultimate to the highest, apophatic path.[22] In extant writings such as *On the Divine Names* and *On the Celestial Heirarchy*, Dionysios offers us a lively and engaging example of the spirit of *theōria physikē* as he explores how God can be known from creation—for as he explains, "truly the visible is the manifest image [*eikōn]* of the invisible" ("Letter Ten," 1117B). Von Balthasar has characterized these writings of Dionysios in a manner that nicely conveys the savor of that authentic love for creation, which Florensky argues is rooted in authentic Christian asceticism:

> One can only with difficulty resist the temptation to quote profusely the theological portrayals by this poet of water, wind and clouds, and particularly of the fragrance of God, the delightful interpretations that go right to the heart of such things as bodily eating and drinking . . . . Particularly noteworthy are his accounts of the spiritual significance of the colours, of the essential properties of the beasts—the might of the ox, the sublime untameability of the lion, the majesty, powerful ascent, swiftness, watchfulness, inventiveness, and sharp-sightedness of the eagle, and so on—[and] of the power of symbolic expression found in the human body and its organs: this gives one an idea of what natural contemplation (*theōria physikē*) can mean in Greek theology.[23]

V

It was, however, a seventh-century Byzantine monk named Maximos who brought this understanding of *theōria physikē*, and the underlying ontology that it entails, to its fulfillment. Although until recently he has been relatively unknown in Western circles, Maximos has long played an influential, foundational role in the Greek East, comparable to the importance of Augustine in the Latin world. Happily, his writings are now beginning to be translated into English.

With Maximos, it becomes even clearer than before that the cosmic *logoi* are neither Platonic forms, nor Augustinian exemplars, nor "ideas" in the mind of God, to which individual embodiments would have to give way, and which would at the same time compromise the darkness and mystery of the divine being. Nor are they Aristotelian essences awaiting their "accidental" instantiation. Rather than essentialist, Maximos's understanding of the *logoi* is intentionalistic and semiotic: They represent God's purpose for each thing and hence the inner meaning of that thing. Every existent being—every person and every animal and every leaf and every stone—has its own *logos*: what it is meant to "say," what the divine Logos is saying *to* it within its own depths, and saying *through* it in its interrelation to other beings. Thus, "to understand them properly," notes David Bradshaw, "requires 'hearing' them, as Moses heard the fire in the burning bush, as part of a discourse uttered by God."[24] Moreover, because the being of these *logoi* is semiotic and intentional, they are able to cohere together naturally into groupings of higher generality without their individuality being subverted or compromised as they would be within a substantialist understanding—just as the verses of a poem yield a higher meaning when taken together in stanzas, sections, or books, and just as the loving acts of a parent elaborate collectively a unitary intention without being reducible either to one another or to the more general intent. Thus, as element of the divine energies discussed earlier, the *logoi* are in Bradshaw's words, "the refracted presence of God in the world, that through which God manifests Himself in His creative act and by which He can be known."[25] Taken together, these *logoi* articulate the eternal Logos in what Maximos saw as a great, Cosmic Liturgy, mirrored by (and mirroring) the Divine Liturgy celebrated in Byzantine temples, that by the time of Maximos had itself for centuries been designed to symbolize the interplay of heaven and earth, as discussed in Chapter 4. For what in the Latin West came to be called the Mass was in the Greek East called the Divine Liturgy, one of whose central features is the joining together of visible and invisible, God and humanity, nature and supernature, nave and sanctuary, *earth and heaven* into a mystical unity that does not cancel difference, but rather enhances it. As Maximos puts it, "God's holy church in itself is a symbol of the sensible world as such, since it possesses the divine sanctuary as heaven and the beauty of the nave as earth. Likewise, the world is a church since it possesses heaven corresponding to a sanctuary, and for a nave it has the adornment [*diakosmēsis*] of the earth.[26]

Just as this unification takes place in the microcosmic Divine Liturgy through the prayerful agency of a priest, so too was this supposed to have taken place in the macrocosmos, the world at large, through the cosmic priesthood of humanity. Indeed, it was just this for which humanity was created: to apprehend and celebrate in the beauty of the cosmos the invisible *logoi* within the visible world. And it is precisely the failure to do this—the failure to exercise a contemplative relation to the visible, to see it inwardly, spiritually, *poetically*—that constitutes the Fall, according to Maximos. Moreover, by seizing upon the visible in a merely sensuous way—approaching it in its outward character according to how it can serve our passions, approaching nature with what Evagrios had called a "demonic" understanding—not just humanity, but nature itself falls into disorder and corruption.

Theōria physikē—or sometimes as with Origen, simply *physikē*—prior to Maximos is largely a station along the way. But with Maximos, it acquires a wholly new status. It is now seen not simply as the means to an end, one stage on the ascetic path, but as an end in itself: both the restoration of paradise, and the realization of our true destiny as human. This is how we once lived on earth, and how we need to be living now. It is the retrieval of how we should live, how we were meant to live: the inception of a life that can be fully lived. It is also the restoration of a mode of knowledge that extends far beyond our life in the midst of the created order, for it concerns our relation to scripture as well as to the Law, both of which must be understood iconically. And once again, it presupposes a notion of redemption, and indeed of the entire path to *theōsis* or divinization, as cosmic, i.e., as connected to the redemption of the world.

So we must practice *theōria physikē* not only to retrieve the paradisiacal, but to exercise that cosmic priesthood that is the truest and highest goal and meaning—the highest *logos*—for humanity itself. For only human beings belong at once to both the visible and invisible orders, and it is within the human element that the nascent unity between the visible and invisible, between nature and supernature, between paradise and world, can be realized and articulated. Drawing deeply upon the theology of the Eastern Church, Maximos sees the Incarnation of the Eternal Logos in a single, particular, *real* human being as laying the foundation for this reconciliation process that is fully cosmic in its dimensions: It is not only for fallen humanity, but for a fallen cosmos that the Incarnation takes place: "the union of our humanity with the divine Logos through the incarnation has renewed the whole of nature."[27] Moreover, just as with Origen, Gregory, and Evagrios, this practice of *theōria physikē* requires preparation: a perfecting of virtue and a purification from the passions, things sought after systematically within a monastic context, but by no means unfamiliar to those who live reflective lives, as the autobiographical accounts of many natural history writers make evident. For it is the passions that make us grasp only the outward, sensuous aspect of nature—seizing only upon its potential for convenience or profit or entertainment—and ignore its inner depths, i.e., the *logoi* that join it to the eternal and invisible. To fail to do this, then, is to approach the visible world with the same kind of clumsy, sensuous, passionate manner as those who read the scriptures in a literalistic and exterior manner—it is to be, as it were, a fundamentalist of the senses! It is, as Maximos draws yet another analogy, to ignore the inner spirit in favor of the rigidity of the law, to be a legalist of the visible. And to phenomenological ears, it cannot help but seem to entail an entrenchment in the "naturalistic attitude" that must be suspended if real knowledge is to be possible.

There is, indeed, beyond this an even higher mode of contemplation according to Maximos, one in which we exercise an even more sublime capacity to enter into the loving, interior unity of the divine being. And *theōria physikē*, the contemplation of nature, does indeed serve as a pathway to this higher contemplation. But it is not *merely* a path, not only a means, nor is it ever superseded or suspended, for it constitutes a crucial aspect our very destiny as human, as not only microcosmic, but as peerless inhabitants of both worlds, of the visible and invisible, and even more as priests of their sacramental reconciliation and communion: "The human person unites the created nature with the uncreated through love . . . showing them to be one and the same through the possession of grace, the whole [creation] wholly interpenetrated by God, and become completely whatever God is, save at the level of being, and receiving to itself the whole of God himself."[28]

VI

Given this view of *theōria physikē,* far from remaining the Nietzschean arch enemy of the earth, the genuine contemplative now appears to be its celebrant and high priest. But important questions remain. How, for example, does this conclusion extend beyond a few extraordinary monastics, scattered about some of the more remote areas of the Eastern Mediterranean, and in a distant past? Yet monasticism itself during the early centuries of Christian history was far from marginal. Rather, many of the most influential figures lived ascetic and monastic lives, and although they often inhabited remote places, their effect on worldly society was widespread and profound. Indeed, the tradition that followed from these five thinkers (and other, similar figures who could have been discussed) formed in fact the main current of thought and spirituality during the first eight hundred years, and it continues as such today in the Byzantine East, where it has been preserved, practiced, and embodied for two millennia. The perpetuation of this contemplative orientation toward nature in the Orthodox East is, of course, largely either unknown in the West, or else written off as a curious remnant of paganism. For example, the kinds of stories told in the West exclusively about St. Francis—tales of consorting and communing with the animals as with friends, and which have led amazed environmentalists to dub him the "patron saint" of ecology—have been told in the East about virtually *all* of the great holy men and women from the desert fathers and mothers of the fourth century to St. Seraphim of Sarov in the nineteenth, and the Elder Paisios of Mt. Athos in the twentieth, a fact that is either overlooked or dismissed as apocryphal in the West. Or if one of the central themes in all of Dostoevsky's major novels is "the moist earth," questioning how our errant passions alienate us not only from God and humanity, but from the very earth itself, this is dismissed—as it is in a recent scholarly study of "Dostoevsky's Religion"—as "pagan pantheism" and "almost pagan earth worship," because its author has decided that it "contradicts some of the historically fundamental aspects of Christianity," i.e., precisely the Western Christianity that he, along with Nietzsche, can then go on to conveniently dismiss.[29]

How, then, was the thread of this tradition lost in the West? Other chapters in this book deal with various aspects of this decline. To varying degrees, it moves forward with Augustine's situating of divine grace outside the order of nature. With the tenth century, Carolingian rejection of the icon as a visible window on the invisible. With the increased insistence that mysticism is not our human birthright, and that contemplative insight must be divinely infused, and then only for an elect few, and only for brief periods. With the scholastic belief that nature and supernature, heaven and earth, are metaphysically separate, as if nature could be understood even provisionally without reference to the divine energies that are always already at play within it. With the final catastrophe of Ockham's nominalism that decisively cut the cord linking the visible and invisible, severing beings from their own depth dimension. And of course, with the consequent valorization of discursive rationality (especially as deployed in modern science and technology) as the only legitimate modality of knowledge.

But neither was *theōria physikē* ever entirely obscured, even in the West. The ninth century Irish monk Eriugena discussed something similar in words drawn from the texts of Maximos, Gregory, et al., that he had been translating, for he was one of the few Latin scholars able to read the Greek texts. St. Bonaventure, follower of St. Francis, also talks about something like the contemplation of nature, but it ends up being reduced to discursivity and the conventions of medieval symbolism. It is revived fleetingly, in Zen-like flashes, in Tauler, Eckhart, and other German mystics. And it is pursued with varying degrees of lucidity and success in German Idealism and English Romanticism by Goethe and Hölderlin and Schelling, by Blake and Wordsworth, Hopkins and Lawrence.

But it is perhaps in American Transcendentalism and its environmental descendents that *theōria physikē* is most impressively re-discovered and retrieved, even (and especially) with regard to its ascetic prerequisites. Thoreau ascetically simplifying his life on Walden Pond. Dillard withdrawn into pilgrimage at Tinker Creek. Muir taking off for the High Sierras with just a jacket, a loaf of bread, and a few tea bags. Monastic fare, indeed. Secular monastics struggling to "see" nature as it is, see it deeply and contemplate its rootedness in the holy. Thus, the great nature writers—who may be the truest founders of modern environmentalism—Thoreau, Emerson, Burroughs, Muir, Leopold, Snyder, Lopez, Berry, Dillard, and many others—could best be read as attempting to retrieve and articulate the contemplative seeing that had become lost in a world that no longer regarded nature as an interface between the visible and invisible. Nor has this literary quest for the sacred in nature abated. Wendell Berry, himself a poet, even designates contemporary nature poetry as a whole as a "secular pilgrimage," in search of "the presence of mystery or divinity in the world."[30]

Yet another poet should be considered here, one who deserves to be included within the more usual canon of environmental poets and writers. For no one in the West during the twentieth century pursued contemplative knowledge, *theōria*, more insistently than did Thomas Merton—in his scholarship, in his ascetic practice, and in his own mode of spiritual environmentalism—and none did more to show that the contemplative life was by no means irrelevant to the earthly, both in his essays and in his poetry. And Merton returned continually to the Greek and Russian East for his inspiration, often writing explicitly of the *theōria physikē* that he had discovered in his readings of St. Maximos the Confessor and the other Greek patristics, even as he confided in his correspondence that he wanted to learn more about it.[31] For example, Merton breaks with scholastic usage as he presents a central insight into *theōria physikē*: "Now the word 'natural' in connection with this kind of contemplation, refers not to its origin but to its object. *Theōria physikē* is contemplation of the divine in nature, not a contemplation of the divine by our natural powers."[32] His friend and biographer, Basil Pennington, stresses the importance of Byzantine spirituality, and especially the practice of natural contemplation, in the development both of Merton's spirituality and his poetics, and writes at length about what he calls "the ever deepening insight into *theōria physikē* that Merton was integrating into his perception of reality."[33]

Hence my claim—which this chapter has merely laid forth—that it has been the partial recovery of this *theōria physikē*, and its correlative ontology of an inner depth to nature, that has served as a spiritual foundation for the rise of modern environmentalism. It would be, then, not by accident that Christian theology reverberates—often quietly, but sometimes, as with Muir and Dillard and Berry, quite explicitly—within the writing of so many of these environmental visionaries. And if this is the case, then although it is our use of Western technology that has pushed the earth to the point of collapse, the sense of nature that has made this possible is itself due to a *spiritual* collapse from which we are yet to recover.

CHAPTER 8

Seeing God in All Things

Nature and Divinity in Maximos, Florensky, and Ibn 'Arabi

> Go to the pine if you want to learn about the pine, or to the bamboo if you want to learn about the bamboo. And in doing so, you must leave your subjective preoccupation with yourself. Otherwise you impose yourself on the object and do not learn. Your poetry issues of its own accord when you and the object have become one—when you have plunged deep enough into the object to see something like a hidden glimmering there.
>
> —*Basho*[1]

> There is only One World, and this is It. What we look on as the sensible world, the finite world of time and space, is nothing but a conglomeration of veils which hide the Real World.[2]
>
> —*Shaikh Ahjad al-Alawi*

I

Why wish to see God in all things? Or what amounts to the same question, why wish to see all things in God? Aren't things in themselves—just the pine and just the bamboo—fine enough, without needing to serve as vehicles for a seemingly extraneous agenda, windows for some monotone view of the divine? But what does it mean to become one with pine or bamboo? And as phenomenology has shown, if it has shown anything at all, the difficulty lies precisely in getting to the things themselves, a process that involves a cleansing of consciousness strikingly parallel to the purification or *katharsis* undertaken in all great traditions of mysticism, each of which has in one way or another emphasized that it is only the "pure in heart" who shall see God in such places as the lilies of the field and the birds of the air, i.e., who can see God in all things. The phenomenological reduction as a path to seeing things, and ascetic purification of the heart as a path to seeing God—these are surely different ways. But perhaps they are not unrelated. As Blake put it, straddling both approaches, "If the doors of perception were cleansed," then we would see everything in its eternal dimension—see that "everything that lives is holy." Perhaps, too, the phenomenological movement of the first half of the twentieth century can be understood as a moment when that great project of recovering the Edenic experience of the world—of seeing the natures of things freshly, catching sight of them as they first emerge into the world: "now, when day is breaking," as Hölderlin had put it, when awakening Nature "begotten out of holy Chaos" "feels herself anew"—when this Edenic impulse actually breaks through, even if momentarily, into the tangle of reasoning that had come to constitute modern philosophy.[3]

This convergence of the phenomenological with the poetic and mystical is embodied in the later thought of Martin Heidegger, who adds to this confluence a powerful rethinking of the historical that gives his work special significance for environmental thought, suggesting that "environmental" problems are not simple "glitches" remediable through technological "fixes." They are, rather, much more deeply rooted, arising from the very manner in which nature is revealed to us. For not just our concepts, but our very experience of nature is radically different from that of pre-modern peoples. In premodern epochs, as Heidegger maintains in the *Zollikon* protocols, "beings were understood as present in and of themselves. For modern experience, something is a being only insofar as I represent it. Modern science rests on the *transformation of the experience [Erfahrung] of the presence of beings* into objectivity."[4] Whereas earlier, beings were experienced as self-emergent, as arising from a hidden abundance, i.e., experienced in the manner once designated by the Greek word *physis*, today we encounter things as objects as we set them over against ourselves as knowing subjects, i.e., not with regard to how they present themselves, but in the mode of their objectivity or representedness. And "this kind of experience of beings," he emphasizes, "has taken place only since Descartes, which is to say only since the time when the emergence of man as a subject was put into effect."[5] Thus, if we are to address successfully the problems of environmental crisis—problems such as habitat loss, destruction of wilderness, depletion of species, and now global climate modification—we will need to do more than tinker with our practices and technologies: We will need to experience nature in a new and different way. But what would this new mode of experience look like? And how would we arrive at it?

Heidegger maintains that nature as a realm of objectivity was itself shaped historically through the onto-theological concept of nature as *ens creatum*, as a realm of created beings whose own being is due to the causal agency of an external, non-created being. At the same time, he defines the contrasting, ancient experience of nature as *physis* through the indwelling of what he calls *die Göttlichen*, gods or divine ones. Nature was then, Heidegger writes in the *Beitrage*, "still *physis*," and thus it served as "the site for the moment of the arrival and dwelling of gods."[6] Both experiences of nature, then, in ancient Greece and in Latin scholasticism, are definable by their respective relations to divinity. Yet with its broken discourses on gods in their arrivals and departures, Heidegger's own philosophical theology is difficult to construe and offers minimal commensurability with any concrete, living mode of spirituality.[7] Nor is it at all clear that the Late Scholastic notion of nature as *ens creatum* is exhaustive, or even representative, of the concept *and experience* of nature as creation either in Judaism, or Christianity, or Islam in their more traditional forms. As we seek alternatives to the objectifying experience of nature that characterizes modernity, Heidegger leaves us without recourse either to the discourse of gods from the ancient world, or to the discourse of creation from the monotheistic traditions nearer to us. Heidegger sees the latter as inextricably bound up with the onto-theological objectification of nature, and the former could be, for us, no more than hints and indicators of some indeterminate reversal of the "flight of the gods" "for which we can do no more than wait in silent anticipation."[8]

Yet the accounts of the great mystics of the past, both distant and recent, in many different times and places, offer us something more: experiences of what Heidegger himself calls "divinely beautiful nature" that possess impressive similarities with one another, while resonating powerfully with us even in a post-modern era. At the same time, they also cohere in unexpected ways with many of the poetic visions of nature (in writers such as Thoreau and Emerson, Dillard and Lopez and Berry) that have shaped the modern environmental movement far more, and far more appropriately, than have the findings of the natural sciences. All, mystics and poets alike, articulate what Basho calls a "hidden glimmering," within the things themselves, of the invisible within the visible—point to the beauty that Heidegger in his elucidation of Hölderlin calls the "pure shining," which takes place when the infinite relation between heaven and earth is disclosed—pursue beauty as what Plato saw to be that singular mode of presence of the eternal whereby it can be experienced within the temporal. Thus, as Basho's injunction suggests, to experience the pine itself would be at the same time, and above all, to encounter within it that hidden glimmering that reveals its rootedness in the invisible, even as we surpass our own subjectivity to join ourselves with the thing itself.

William James, whom Heidegger dismisses in his lectures on the phenomenology of religion as a mere psychologist, argues on epistemological grounds that mystical experience is the "root and center" of all religious experience. Beyond this, James maintains that mystical consciousness not only subverts, but "absolutely overthrows" the hegemony of everyday experience, first of all because it is apodictically compelling for the mystic, who having experienced both mystical and non-mystical states, is alone in a position to choose between their relative veracity. But perhaps more important, mystical experience presents itself as a fulfillment of that toward which ordinary experience always tends, "point[ing] in directions to which the religious sentiments even of non-mystical men incline."[9] Or as many classical figures have maintained, mystical experience articulates what comes to be realized as having been there all along, even if unnoticed or inchoate or tacit: the unsayable within the saying, the residual element within experience that serves as its possibility. And as James notes, the most common form of mystical experience is that moment in which we suddenly and finally "see," deeply comprehend, the truth of something we have known all along in a conventional or merely discursive manner. Perhaps this is also that dimension within our experience of nature, often tacit or hidden altogether, that anchors the natural itself, keeping it from being nothing more than our own objectification of it, our own possibilities for comprehending and mastering and using it.

So it is to certain figures within the great traditions of mysticism that I shall now turn for directives concerning our experience of nature and thus our relation to it—rather than to Heidegger, whose familiarity with the literature of classical mysticism, which he often characterized as "irrationalism," may have been much more limited than is commonly supposed. Three figures will be briefly considered: one of them a voice from ancient Constantinople and the Byzantine lands of the seventh century; one from Andalusia and other great centers of thirteenth-century Islamic civilization such as Fez, Cairo, Mecca, Baghdad, and Damascus; and one, contemporaneous with Heidegger's earlier writings, speaking from within that last glimmer of the ancient world, during the inundation of Russian spirituality that took place with the Bolshevik triumphalism of the everyday and ordinary in the early twentieth century. In each case, the focus will be upon the numinous, theophanic point of union: between the one and the many, between the visible and the invisible, between time and eternity, between heaven and earth. After these three synopses, we will, in an even more summary manner, consider the prerequisites for these kinds of experiences of nature and cosmos, as well as the implications they might suggest for seeing God in all things.

II

Although he has exerted a powerful, if often subterranean influence in the West upon thinkers from John Scotus Eriugena to Jean-Luc Marion, St. Maximos the Confessor has been widely appreciated only in the Greek East, where since the time of his contemporaries he has exerted a formative influence, perhaps comparable to that of St. Augustine in the Latin West. Whereas much of Augustine's thought revolves around questions of faith and knowledge, Maximos sustains a focus upon the process of *theōsis* or divinization, which he relates not just to humanity but to the entire cosmos. By the time of Maximos in the seventh century, the path of spiritual transformation had already been well-marked in the East, where it was seen to consist first in *katharsis* or purification, second in *theōria*—illumination, contemplation, or "seeing"—and finally in *theōsis* or divinization, union with God that at the same time draws together God and cosmos. For the soul to see clearly and to see deeply, it must first be purified, a principle that is ubiquitous in the literature of world mysticism and that was articulated in modern times by William Blake in *The Marriage of Heaven and Hell*: Since we have "closed [ourselves] up till [we] see all things thro' the narrow chinks of [our] cavern," if we are to be raised up "into a perception of the infinite," and indeed for the "whole creation [to] appear infinite and holy, whereas it now appears finite & corrupt," then "the doors of perception [must be] cleansed." The highest capacity of the soul—the *nous* for Maximos, who retains much of the Platonic lexicon, although in a much transformed mode—must be freed and purified from its entanglements both with passions and with discursive thought, if it is to become aware of the infinite within the finite, the visible within the invisible, the eternal within the temporal. Maximos characterizes this contemplation of nature and cosmos as *theōria physikē*, a term he inherited from an already venerable tradition of mystical *praxis*, associated with such figures as Origen of Alexandria, Evagrios the Solitary, and St. Gregory of Nyssa. Moreover, he argues that this noetic practice of *theōria* also exercises the same faculties of seeing the invisible within the visible that should be employed in reading scripture, a focus upon the inner meaning as opposed to a "fundamentalism" either of the senses or of the scriptural letter.

The difficulty of "seeing" nature, unencumbered by the worldly blinders endemic to the *civitas*, is a problem that Maximos and his predecessors share with modern environmentalists such as Emerson, Thoreau, and especially Annie Dillard, who devotes the key chapter of her account of a pilgrimage through nature toward the holy, to the problem of "seeing"—which turns out to be not just the means but the end for her pilgrimage as well. For Maximos, what is truly to be "seen" throughout nature are the divine *logoi*, each unique and specific to every creature. Diverging from classical Greek epistemology—for which legitimate knowledge must have the universal as its object, with the singular object as no more than a partial instantiation of the universal—Maximos maintains that the inner principle or *logos* of every being is not only intelligible to noetic apprehension, but that it is what above all *is* intelligible. Thus, these singular *logoi* are by no means exemplars in the Augustinian sense, which are indeed Platonic and non-individual, and are susceptible to Nietzsche's charge that the eternal and invisible drains the temporal and the visible of their weight and meaning. The *logoi* of Maximos, rather, instill and infuse the visible with meaning, continually restoring beings to the wellsprings of their own interiority, and resisting the maze of thoughts and passions we ordinarily impose upon our knowledge of them, whether that imposition is the objectivity of modern scientific cognition or the commodity status of the ancient bazaar. Rather than exemplars or ideas, whose being is universal, the *logoi* are words, divine "sayings," the divine and eternal words spoken from all eternity, and rendered incarnate in the appointed time and place. They are in each case what God "has to say" in and through particular things.

As divine words, the *logoi* can be understood within lesser and greater contexts, just as each line of a poem can be understood against the background of other lines and stanzas, and the poem itself can be grasped as part of the entire opus of the author, without ever losing its individual standing. Or rather, that individual standing is increasingly enriched as it is progressively contextualized within larger unities, until together the *logoi* are seen as expressive of a single, eternal *Logos*. This is the understanding that gives rise to the notion of the "book of nature," a vision of nature that first arises not within the milieu of Newton and Galileo, but in the Egyptian desert of the fourth century, when St. Anthony the Great could reply to visitors that he needed no books because he could read the book of nature everywhere around him. But the lexicon to which he refers lies not in the mathematical projections of modern natural science, but in the inner principles of things that can only be comprehended noetically, not discursively: reading nature for its inner meaning, just as we look toward the inner meaning in scripture. Maximos thus affirms, along with the Sufi master al-Alawi cited earlier, the unity of the visible and invisible. As Maximos puts it, almost identically to al-Alawi more than a millennium later: "There is but one world and it is not divided by its parts. On the contrary . . . the spiritual world in its totality is manifest in the totality of the perceptible world, mystically expressed in symbolic forms for those who are capable of seeing this. And the perceptible world in its entirety is secretly fathomable by the spiritual world. . . ."[10] This unified apprehension of the perceptible and the spiritual in nature is what allows humanity to assume the cosmic priesthood intended for it, consecrating and sacramentalizing nature through our contemplation (*theōria*) of its divine inner principles, returning "to our [own] nature the very beauty for which it was created in the beginning," while allowing us to see the inner beauty of nature as a whole to which the senses alone are impervious.[11] Observing the one *Logos* running throughout the *kosmos*, we "display the grace of God effective to deify the universe."[12] It is thus the task of each one of the faithful, not only of monastics, to carry out this contemplative work so that the divine, in Maximos's words "might be *through all things and in all things* (Eph. 4:6), contemplated as the whole reality proportionately in each individual creature . . . and in the universe altogether, just as the soul naturally indwells both the whole of the body and each individual parts without diminishing itself."[13]

Thus, in such a divinized *kosmos*, God dwells within each being just as the soul dwells within each part of the body, without thereby being in any way reduced to it, for all possess, in Maximos's words, "a kind of innate power clearly proclaiming God's presence in all things."[14]

III

Muhyî al-Dîn Ibn al-'Arabi is widely considered to be not just the greatest of Sufi philosophers, but arguably the greatest philosopher of Islam. Yet in the West his thought is eclipsed by secondary thinkers such as Ibn Rushd (Averroes) who had been more influential in Latin medieval circles, while in many parts of the Islamic world his thought is often seen as questionable, if not heretical. From his exceptionally rich and extensive writings, we can focus here upon only a single notion, and this in a very abbreviated way. Remarkably similar to the individual *logoi* of Maximos is the notion of the *wajh* or "face." Every being possesses its own specific, individual, or as it is sometimes rendered "private" face that is unique to it. And it is this private or specific face (*wajh al-khâss*), rather than its materiality, that forms its very uniqueness and individuality. Ibn 'Arabi understands this private face as the face that each being turns toward the divine face, toward its own Lord, for whose worship that being was uniquely created. "Every creature," he maintains, "has a private face to its Lord."[15] "The 'face' of anything is its essence . . . [that is,] the locus of turning toward God, apart from the other directions. It is the most magnificent direction."[16] And reciprocally, the Lord of that being is the face that the One God, who lies beyond all experience, turns toward that individual being alone: "God the Most High has a private face in every existent. He imparts to him from it whatever He wants, of that which is not for any of the other faces, and through that [private] face each existent is needy of Him."[17] Thus, God is manifest in every being, all the while remaining at the same time beyond all. This ontological relation of each being to its Lord allows Ibn 'Arabi to undertake a remarkable epistemological turn, essentially reversing the cognitive standing of the first two parts of Plato's divided line, and maintaining that imagination is higher than ordinary, empirical experience. For if my very being is metaphysically oriented in this most singular way to the Lord that is mine alone, then imagination, visions, even dreams will portray that singularity better than the confused impressions of external experience, and for this reason Henri Corbin (who before becoming a scholar of Islamic mysticism, completed the first translation of Heidegger into French) has coined the term "imaginal" to designate the realm that Ibn 'Arabi intends, in order to distinguish it from the mere subjectivity implied in the term "imaginary."

Thus, every being is both a manifestation and a veil of the divine, and the highest understanding of beings must be an unveiling that will make manifest the aspect of God that it uniquely presents: "The universe is a place of signs, and every reality in it is a sign that leads to a divine reality, upon which it is founded in coming into existence, and to which it will return when withdrawn [from existence]."[18] Hence, as in Maximos, we find prescribed by Ibn 'Arabi a knowing that is at the same time phenomenological (a "seeing") and hermeneutical (an exegesis of the ontological depths of the thing itself). He calls this process *ta'wil*, and here, too, the symbolic exegesis is to be applied to the natural world, to scripture, and to religious ritual: In each case, we must proceed from the external and literal toward spiritual interiority. And especially with regard to the realm of nature, we are able through this unveiling-seeing to arrive at the Divine Wisdom or Sophia, which Corbin characterizes as the "Creative Feminine" principle underlying all creation.[19] But two other features also characterize this unveiling: first the necessity of an ascetic purification of the soul, common to all authentic mystical paths, and of which Sufism is rich in resources. And second, the phenomenon of love, for it is not the cold glance of the investigator, but the warm heart of the lover that properly unveils and makes manifest: "it is impossible to adore a being without conceiving the Godhead in that being . . . [for] a being does not truly love anyone other than his Creator."[20] Not surprisingly, the locus of knowing is not in the head, but within the vastness of the heart—understood not as a seat of passion or emotion, but as is the norm in traditional cultures, as the center of a knowing far deeper than discursive rationality. Finally, with Ibn 'Arabi as with Maximos, it is through our experience of beauty that we apprehend the divine face within the private face of each being: "If you love a being for his beauty, you love none other than God, for He is the beautiful being."[21] Proceeding on this path of love and contemplation, the soul "contemplates God in every being, but [does so] thanks to a gaze which is the divine gaze itself."[22]

IV

Pavel Florensky, murdered by the KGB in 1937, was arguably Russia's greatest philosopher, and probably Europe's most dazzling intellect since Leibniz. His *Pillar and Ground of Truth*, published in 1914, and recently translated into English, possesses a philosophical power and scope that invites comparison with Heidegger's *Being and Time*. Although he was married with a family, and spent much of his adult life as a university professor and conducting scientific research, Florensky not only possessed the spirit of a monk, but in ways that were still possible in pre-revolutionary Russia, lived much the life of a monk. From Florensky's extraordinarily rich work, we will focus here only on the relation between asceticism and the love of nature. For Florensky, ascetism should be understood not just in the strictest sense of fasting, vigils, and so on, but more broadly as it is expressed in the Russian term *podvig*, as a self-overcoming, and indeed an *overcoming of self*. Of particular interest for us is his distinction between two kinds of asceticism, modes that he argues are complete opposites to one another: one of them looks at nature, and rejects its outward beauty as illusory, as a mere façade for what is inwardly corrupt and vile, and thus a realm from which we should flee. This is the ascetic spirit that Nietzsche subjected to merciless criticism, and justly so. In contrast, the other, genuine mode of asceticism seeks to go beyond the outward beauty not toward another world, but deeper into this one: to see the surface beauty as insignificant *only* in comparison to a deeper beauty of nature, and to employ the ascetic impulse not to flee from nature but to reveal a greater beauty than the mere "charm" that ancient figures such as Hesiod (in *his Works and Days*), Socrates (in the *Phaedrus*), and Virgil (in his *Georgics*) appreciated, without having the means of going farther. The first of these is an asceticism of revulsion and escape, while the other is an asceticism of love for, and joy in, creation. Florensky argues that it is this latter asceticism, first typified by the Desert Fathers (and Mothers) of the Fourth Century and later, that for the first time discovered and articulated the love for nature upon which modern environmentalism draws for its inspiration. "Thus the goal of the [genuine] ascetic's strivings," he argues, "is to perceive all of creation in its original triumphant beauty."[23] It is the purified heart that can overflow with love for all creation, "when what is revealed to it is that side of every creature which is worthy of total love and which is therefore the eternal and holy side of every creature"—a vision with powerful correlates in Maximos's *logoi* and Ibn 'Arabi's "private face."[24]

As with Ibn 'Arabi, this authentic mysticism seeks to purify the heart so that it can see clearly the invisible within the visible, the holy within nature. As in Heidegger, the holy in nature is, for Florensky, characterized by its divinely beautiful character. And again as with Ibn 'Arabi, Florensky sees the experience of this holy beauty of nature as an apprehension of the Divine Wisdom, "the perception by ascetics of the eternal roots of all creation by which creation is anchored in God," a "manifestation, imperceptible for others, of the heavenly in the earthly."[25] Here, too, this higher knowing of nature not only presupposes, but precisely *is* an act of love, a union of knower and known. Discarding the assertive boundaries of the monadic ego—as "love takes the monad out of itself" and identifies it with God's own self-emptying love for creation—the ascetic's heart is deified, joined with the divine love: "the true I of a deified person, his 'heart,' is precisely God's Love, just as the essence of Divinity is intra-Trinitarian Love."[26]

V

Surveying these three figures, widely separated in time, place, and culture, we find a remarkably coherent vision of how the mystic's experience of nature fulfills abundantly the environmentalist agenda that we can derive from Heidegger. With more time, compelling examples in the literature of Judaism, Taoism, Buddhism could easily have been added. But how did Heidegger himself fail to see this possibility? I believe the answer lies in certain developments in Latin Christianity that began in the first millennium and that culminated in Late Scholasticism—a development that shaped Heidegger's view not only of Christianity, but of religion as such. Prefigured in Augustine, further developed in the ninth century *Librii Carolini*, and finalized in Scholastic philosophy, the Western Church came to see the experience of the divine energies within the world—the mystical experience as such—to be something exceptional and in need of being supernaturally "infused," rather than as the birthright of all humanity. Or rather, since Latin theology had long ignored or rejected the distinction between the divine essence (*ousia*) and divine energies (*energeiai*), mystical experience in the West increasingly bypasses creation altogether, becoming localized in rarified and extraordinary visions such as those of St. Teresa of Ávila. But what if James is correct in his belief that mystical experience lies at the heart of all genuine religion, not just in its origins but in its living practice? Then the Western eclipse of the mystical would represent a severance from the source of religious life itself, leaving only ecclesiastical authority, the abstractions of so-called "natural" theology, and the will and emotionality of "faith" as the supports for spirituality, despite the fact that by themselves, they are really capable of supporting only dogma and obedience in all but the learned, who can satisfy themselves with splendid abstractions.

Put simply, there are strong reasons for believing that it is primarily, and perhaps exclusively, the European West during the last millennium or so that has had any serious difficulty in "seeing God in all things." (This is, perhaps, the principal reason why prior to the eleventh century, it never occurred to anyone, even in the West, to employ discursive rationality to "prove" in any formal manner that God existed.) Nor, I believe, is this unrelated to the fact that it is within the realm of Western cultural dominance that nature has been so radically despoiled. Failing to see this, Heidegger foresees the possibility of what he calls "saving" the earth as bound up with occult and nebulous possibilities, for which we could do little more than prepare in some very tenuous manner anyway. But if the reunion of earth with its heaven is a possibility that lies at the mystical heart of all great religious traditions (vital, existing traditions that hold genuine possibilities for actual people) then a retrieval of the mystical elements from within these traditions—and a de-construction of their Western and modernist distortions, of which both "progressive" revisionism and "reactionary" fundamentalism are equally symptomatic—this de-construction and this retrieval would constitute a project that was both tenable and urgent.

CHAPTER

9

The Glory of God Hidden in Creation

Eastern Views of Nature in Fyodor Dostoevsky and St. Isaac the Syrian

> Just as we have two bodily eyes, so we have two spiritual eyes, and each has its own way of seeing. With one we see the glory of God hidden in creatures; with the other we contemplate the glory of God's holy nature when he deigns to give us access to the mysteries.
>
> —*St. Isaac the Syrian*[1]

> What love served to initiate the creation of the world!
>
> —*St. Isaac the Syrian*[2]

Every field has its canonical works, texts exerting such great influence that their conclusions are accepted rather uncritically. And as already noted, one of the few indisputably canonical texts in environmental thought—comparable in influence perhaps only to Also Leopold's "The Land Ethic"—is surely the seminal article by the Protestant historian of medieval technologies, Lynn White Jr., published in 1967 in the AAAS journal *Science*, and entitled simply, "The Historical Roots of Our Ecologic Crisis."

By way of review: Its author indicts Christianity for generating our present environmental problems, arguing that the "victory of Christianity over paganism"—"the greatest psychic revolution in the history of our culture"—entailed a transition from being "part of nature" to being an "exploiter of nature," leading to a "ruthlessness toward nature" that is characteristic of modern technology.[3] White proceeds from his research into changes in medieval technologies in the Latin West (outlined earlier in Chapter 6) connecting these changes to a certain "Christian arrogance toward nature."[4] He goes on to relate this "exploitative attitude" to a desacralization of nature, anticipating what Max Weber had called the "disenchantment of nature," although White situates the tipping point for this transition not in modernity, as did Weber, but in the theology of the Latin Middle Ages, which he believes regarded any notion of "spirit in nature" as "idolatrous."[5]

But as we have also seen, discussions of White's thesis rarely take note of the fact that he repeatedly and explicitly exempts the Christianity of the Greek East from his critique. White praises Byzantine Christianity for its contemplative and richly aesthetic relation to nature, in contrast to the calculative and voluntaristic predispositions of the Latin West, where the more aggressive and "violent" technologies that concern him (like the deep-furrow plow) were increasingly developed and deployed.[6]

Now, having documented what he believes has gone wrong, White wishes also to suggest an "alternative Christian view of nature and man's relation to it," maintaining that "since the roots of our trouble are so largely religious, the remedy must also be essentially religious, whether we call it that or not." Surprisingly, however, rather than looking to the Byzantine East for what he calls "the remedy," as would seem to follow from his argument, perhaps looking into it as the path not taken by the Western Church, he looks instead to the culturally exceptional figure of St. Francis of Assisi, openly admitting that, to the West, St. Francis seems "heretical," a choice made even more ironic by White's apparent lack of awareness that within a few generations, it would be certain members of the Franciscan Order, notably John Duns Scotus and William of Ockham, who would serve historically as the pious advocates of a radically desacralized nature—nature swept clean of divine essence or archetypes, and thus rendered as an opaque end-product of divine volition, a simple *factum, ens creatum,* but surely not *natura naturans*—advancing this new characterization of nature in a way that actually *did* lay the foundations for modern science and technology. Nor does White seem to be aware that Francis's friendly relations with animals had long been a hagiographical commonplace in the Christian East, from the earliest monastics of the Egyptian desert to present-day monks on Mount Athos: St. Gerasimos with his helpful lion, St. Seraphim with his cooperative bear, and Elder Paisios with his guileless snakes effortlessly join company with St. Francis and his penitent wolf. ("The humble man," writes St. Isaac of Syrian, one of Dostoevsky's favorite authors, "approaches ravening beasts, and when their gaze rests upon him, their wildness is tamed. They come up to him as to their master, wag their heads and tails and lick his hands and feet, for they smell coming from him that same scent that exhaled from Adam before the fall. . . ."[7]) Even more surprisingly, after insightfully contrasting the profound differences between Eastern and Western views, rather than regarding the Greek East as pursuing the more viable course, or even as preserving an originally salutary relation to nature that had become lost to the West, White unabashedly begs the question: "The implications of *Christianity* for the conquest of nature would *emerge more easily* in the Western atmosphere," he notes glibly.[8] Pursuing the same logic, one might just as well say of two siblings, the one a firefighter and the other an arsonist, that the second demonstrates "more easily" the essentially incendiary character of the family. If there is, in fact, a historically divergent branch of Christianity whose relation to nature is arguably more salutary, might not the environmental implications of Christianity "emerge more easily" *there*?

The same blindness to the mere possibility that authentic Christian spirituality can not only embrace the holiness of nature, but even venerate it—that "spirit in nature" is far from alien to authentic Christianity—is even more egregious in Western scholarship on the Russian author Fyodor Dostoevsky, himself one of the most important and articulate of Eastern Orthodox thinkers, and in what must come as a surprise to Western readers who approach him as a proto-existentialist or as a dark psychologist, a great poet of the glory of God hidden in creation. Commentators have, in fact, typically done to the Russian novelist just what White does to Byzantine Christianity generally. Dostoevsky presents throughout his novels an analysis of the human soul in which the corruption of the heart is connected to an alienation from nature, and in which redemption is often—as in the case of Raskolnikov—heralded by a restored relation to nature, as is readily apparent in Dostoevsky's last and greatest novels, through characters literally kissing the earth and watering it with their tears of contrition: *Crime and Punishment*, *Demons*, and *The Brothers Karamazov*. Human wrongdoing here is, in an important sense, a sin against nature, a crime against creation, an offense against the birds of the air and even against the earth itself. Perhaps his greatest character, the Staretz (or "Elder") Zosima—who Dostoevsky saw as the "good man" he had long been struggling to present in his fiction, articulating what after much spiritual struggle of his own, he had come to regard as his own highest truths—presents what is arguably the most powerful vision of a compassionate, spiritualized, indeed reverent relation between humanity and nature in all of literature, even when figures such as Thoreau and Muir are included in the pool. Nevertheless, although many Russian critics have very different appraisals, in the secondary literature of Western Europe and America, Dostoevsky's love for nature, and especially his veneration of the earth, are commonly taken as evidence that he could not really have been much of a Christian at all, but some kind of retrograde pagan instead. For of course it is well known that Christianity is inherently hostile to nature. This is, to speak plainly, simply the logic of ancient prejudice and modern bigotry.

Steven Cassedy, for example, in his definitive-sounding *Dostoevsky's Religion*, demonstrates a dazzling incomprehension of Orthodox spirituality. Listing the many characters Dostoevsky identifies with the phrase "kiss the earth," he concludes from this not that Orthodox Christianity has alternative attitudes toward nature, but that "none of these figures is what you might call a Christian in any traditional sense of the word."[9] Of Zosima's sense of the divine energies in creation, the author proclaims as if it were a reproach: "the closer we get to what is meant to pass for Christianity in Zosima and his disciple Alesha [*sic*], the closer we get to . . . the earth."[10] Dostoevsky's many references to the earth throughout his writings are dismissed without argument as "pagan pantheism" and "almost pagan earth worship" merely because they "contradict" what the author thinks are "some of the historically fundamental aspects of Christianity."[11] The circularity here is dizzying! Why look everywhere else than to what is closest: the possibility that alongside its Western variant, the Christian tradition is represented by at least one very different modality—and indeed, one that claims unbroken continuity with Christianity's earliest forms—one that not only allows for the possibility of nature as divine revelation, but that emphatically insists upon it? Parallel to Cassedy is the German scholar Rudolf Neuhäuser, writing in *Dostoevsky Studies*, who flatly states that "love of the earth and joy in being alive and being part of earthly life . . . contradict the traditional view of the world in Christian writing as a 'vale of tears,' full of temptation and sin."[12] If the stereotype is threatened, it seems, so much the worse for the reality![13]

But for those who care to look, the reality is quite the opposite. There is, for example, St. Isaac the Syrian, a seventh-century monk who is mentioned by name several times in *Karamozov*; to whose *Ascetical Discourses* (which had been recently translated into

Slavoinc) Dostoevsky is said to have returned to again and again; and who likely inspired many of the ideas expressed by Dostoevsky's figure of Zosima. St Isaac maintained that God is love, pure "immeasurable" love that is expressed in God's creation of nature itself, and in which we human beings are called to participate through our own self-effacing love for all created things. St Isaac portrays this love for all creation through a narrative that has been immensely influential in Orthodox spirituality and beyond:

> An elder was once asked, "What is a compassionate heart?" He replied:
>
> "It is a heart on fire for the whole of creation, for humanity, for the birds, for the animals, for demons and for all that exists. At the recollection and at the sight of them such a person's eyes overflow with tears owing to the vehemence of the compassion which grips his heart; as a result of his deep mercy his heart shrinks and cannot bear to hear or look on any injury or the slightest suffering of anything in creation.
>
> "This is why he constantly offers up prayers full of tears, even for the irrational animals and for the enemies of truth, even for those who harm him, so that they may be protected and find mercy.
>
> "He even prays for the reptiles as a result of the great compassion which is poured out beyond measure—after the likeness of God—in his heart."[14]

It has often been noted that St. Isaac of Syria had a strong and decisive influence upon Dostoevsky's thought, especially regarding the relation between God and nature, and its meaning for humanity.[15] But apart from the specifics of this influence, congruities with St. Isaac's writings will be helpful here not only in clarifying Dostoevsky's views, but in demonstrating how solidly Dostoevsky stands within the ancient patristic tradition concerning the relations of God, creation, and humanity, rather than representing some modern form of "pantheism" or "nature mysticism." Taken together, their views can cast light upon questions to which this study has returned repeatedly from different directions. What is the congenial and appreciative understanding of nature and our relation to it that typifies Orthodox Christianity, and putatively original Christianity as such? And can Western Christianity, and perhaps other traditions as well, learn something significant from them?

Dostoevsky, especially, can serve as a particularly helpful point of reference in this pursuit, because many Western audiences are already familiar with his work, although often the acquaintance is skewed through employing approaches (for example "psychological" or "existentialist") that are alien to Dostoevsky's concerns. And the aphoristic charm of St. Isaac's sayings has already lent them a strong appeal to modern readers in the West.

Distinctive elements in the Byzantine and Russian relations to nature can here be summarized in seven points, none of which are usually acknowledged either in discussions of Dostoevsky, nor in reflections on Christianity and environmental thought overall. These are important not only in furthering our understanding of environmental issues, but in better understanding Dostoevsky himself. And this, in turn, will help us better situate Western thought and the point at which it has arisen. For Dostoevsky vies with Nietzsche as the great diagnostician of Western nihilism. It forms a central concern in all his important novels. Yet despite the similarity of his diagnosis of nihilism to Nietzsche's, his understanding of the etiology differed in important ways. While Nietzsche saw Western modernity as the culmination of Christian civilization, Dostoevsky saw it as a default from the Ancient Christianity that he believed had declined in the West since the Dark Ages, and had eventually become a rebellion against it. Nietzsche's Zarathustra urges us to be true to the earth, yet understands nature in a way that differs little from the positivism that the young Nietzsche had embraced—a metaphysic just as materialistic, with nature quite as subject to human will, as was the case with Marx, even if Nietzsche (who, as a good German, loved to hike outside) had a keener eye for the fine points of nature than did the city-loving, countryside-despising Marx. Dostoevsky's characters, in contrast, truly revere the earth, kiss it and water it with their tears, admit to it its own darkness, mystery, and holy integrity. Perhaps, as we shall see, an important difference lies in the respective appraisals of humility in the two thinkers. Berdyaev wrote insightfully that Dostoevsky knew everything that Nietzsche knew, but that at the same time Dostoevsky knew something that Nietzsche did not know.[16] Contemporary Western thought has taken Nietzsche's understanding of nihilism and the way beyond it as normative. It will be the burden of this chapter to suggest that perhaps we should seriously consider whether Dostoevsky's understanding offers more helpful insights, especially in view of the impasse at which we have arrived in our understanding of nature, as well the inability of Nietzsche (or his contemporary successors, many of whom have regressed into the nihilism Nietzsche abhorred) to address it.

First, like all representative thinkers of the Christian East, Dostoevsky draws heavily upon the Byzantine distinction between divine substance or essence (*ousia*) and the divine activities or energies (*energeia*), a difference that was never well understood in the Latin West. Going back to the fourth century Cappadocians, but finding their final elaboration in St. Gregory Palamas in the fifteenth century, these concepts differentiate the divine *essence* (radically transcendent, eternally unknowable, and forever unapproachable) from the divine *energies*, which are immanent throughout the created world, infusing and saturating and animating all of nature, and can be encountered and experienced as such. This distinction is therefore especially important for providing the basis for a powerfully mystical orientation to Byzantine spirituality, allowing it to affirm the *experience* of the uncreated energies of God in creation while preserving the radical mystery and transcendence of the divine essence.

Thus, when their eyes have been opened through their attainment of some measure of humility—"humility runs in advance of grace," teaches St. Isaac—Dostoevsky's characters begin to see God in every aspect of creation, without endorsing any pantheistic reduction of divinity to nature.[17] "Everything is a mystery, my friend," explains Makar to young Arkady, "there is God's mystery in everything. Every tree, every blade of grass contains this same mystery. Whether it's a small bird singing or the whole host of stars shining in the sky at night—it's all one mystery, the same one."[18] The peasant Makar points here not to an intellectual puzzle, for which he would care nothing, but to the mystery of divine energies at work all around him and which he knows from his own experience, as his subsequent narrative tells us. And in his "Econium" of praise to St. Isaac, the great Greek iconographer and theologian Photios Kontoglou describes the noetic vision of the divine energies experienced by the humble and purified heart: "He will see another world, and he will hear another harmony which heretofore his calloused heart did not even expect. . . . The earth and the creatures of this world, which all of us see, will then take on an indescribable appearance and an unfading beauty."[19]

Second, as a corollary, there is the Eastern teaching of the *logoi* of creation, also expressed by the early Cappadocian thinkers, but most clearly stated by St. Maximos the Confessor. Every being—every blade of grass and cloud and animal—possesses its own, individual, unrepeatable *logos*, the specific "thought-will" of God that constitutes its being and meaning and identity, while at the same time mirroring the eternal Logos of God. Thus, the Staretz Zosima can assert in *The Brothers Karamazov*, "we have been granted a secret, mysterious sense of our living bond with the other world, with the higher heavenly world, and the roots of our thoughts and feelings are not here but in other worlds . . . God took seeds from other worlds and sowed them on this earth, and raised up his garden . . . but it lives and grows only through its sense of being in touch with other mysterious worlds; if this sense is weakened or destroyed in you." Zosima continues—diagnosing, in effect, the affliction that assails all of Dostoevsky's nihilistic characters, such as the Underground Man, Stavrogin, Raskolnikov, and Ivan—if this sense of contact is weakened, "that which has grown up in you dies. Then you become indifferent to life, and even come to hate it. So I think."[20]

This correlation between the divine Logos and the *logoi* of created things—between the visible and invisible, heaven and earth—is also given powerful expression in the writings of St. Isaac of Syria. "The first book given by God to rational beings," he argues, glossing St. Anthony's understanding of creation as a book teaching divine things, "was the nature of created things. But the instruction set down in writing [i.e., the scriptures] was added [only] after the transgression." And even now, he exhorts, "ask from nature a true witness concerning yourself and you will not go astray. Yet even if you turn aside from thence [i.e., from the *logoi* or meaning found in creation], learn from that second witness [i.e., scripture] not to go astray."[21] And later, in Homily sixty-two St. Isaac clarifies this possibility of learning heavenly lessons from earthly things with a correlation that is both spiritual and ontological: "Does not nature itself instruct us concerning these things? And what heavenly thing is there, no likeness of which can be found in our nature that enables us to speak of it? For behold, even our Lord [for example, in his parables and in the Sermon on the Mount] confirmed all spiritual things by examples from the things of nature and in this manner instilled their power into our souls."[22] This correlation between the visible and the invisible, between heaven and earth, is ontologically grounded on the Orthodox teaching concerning the divine energies, as well as that concerning the *logoi* of creation. The power of this correlation, based ultimately upon the Incarnation, whereby divinity becomes human, heaven and earth are joined together, even leads St. Isaac to speak of a "mingling" between God and creation: "the world has become mingled with God, and creation and Creator have become one!"[23] Against teachings such as these, the claims of Dostoevsky's critics that his looking toward nature for instruction concerning God is somehow pagan or pre-Christian appear uninformed at best.

Third, Eastern Christianity affirms the iconicity of creation: That is, it emphasizes the belief, and the corresponding sensibility, that creation is itself divine revelation, God's first revelation of Himself, presented long before the writing of scriptures. In his "Sermon on the Mount," and in many of his parables, Jesus asks his hearers to look upon the birds of the air, the mustard seed, the lilies of the field, as manifestations of a divine order, as windows upon the workings of God in the world. And just as the sacred art of traditional Christianity—unlike the merely religious art of the Renaissance and its successors—should be taken as presentational rather than representational (in German, as *Darstellung*, rather than *Vorstellung*) so creation can be apprehended iconically, not as mimetically depicting the divine (as Latin, allegorical interpreters would see it) but as offering us windows opening up *upon* the divine, sacramental apertures through which those whose souls have been cleansed could see the invisible within the visible. Accordingly, Zosima exhorts: "Love all of God's creation, both the whole of it and every grain of sand. Love every leaf, every ray of God's light. Love animals, love plants, love each thing. If you love each thing, you will perceive the mystery of God in things. Once you have perceived it, you will begin tirelessly to perceive more and more of it every day."[24] And if Zosima's zeal for seeing God in the natural world around him seems rather exuberant, it is surely excelled by that of St. Isaac the Syrian, who describes (once the mind or *nous* is rendered quiet) a state beyond prayer, which he compares to inebriation:

> When someone reaches insights into creation on the path of his ascetic life, then he is raised up above having prayer set for him within a boundary . . . From here onwards he finds the senses continuously still and the thoughts bound fast with the bonds of wonder; he is continually filled with a vision replete with the praise that takes place without the tongue's movement . . . This in truth is the state of cessation above prayer when he *remains continually in amazement at God's work of*

> *creation—like people who are crazed by wine*, for this is "the wine that causes a person's heart to rejoice."[25]

Fourth and fifth, we may list the correlative notions of cosmic fall and cosmic redemption. In traditional Christianity, and still today in Eastern Orthodoxy, the fall of humanity has been seen as leading to a corruption of all nature, eerily anticipating our contemporary realization of how human vice—avarice and intemperance and pride—has brought about the downfall of the natural order itself. This is an ancient thought that is uncannily recapitulated in modern experience. But as a corollary, the redemption inaugurated by Christ has always been seen as a redemption not just of humanity, but of all creation: God has Himself entered into creation in a new way, allowing it to be restored, transfigured, and deified through the human possibility—for those whose hearts have been purified—to serve as a sanctifying priesthood of nature, apprehending and venerating the Edenic roots of creation, and thereby offering up creation to the Creator as a consecrated sacrifice. Indeed, Byzantine theology sees the uniqueness of humanity not in its power or knowledge, but in the fact that human beings alone are dwellers in two worlds, both the visible world and the invisible world, and hence it is our charge to draw them into unity. In Dostoevsky's writing, when both Zosima and the so-called Ridiculous Man of a later story discover that we are already living in paradise, if only we could perceive it, they are simply giving modern voice to an ancient truth.

Beyond this, the natural order must itself be seen as a cosmic liturgy that joins together with humanity in this priestly chorus. For example, in one of the four great Orthodox liturgies, the ancient Liturgy of St. James, just prior to the consecration of the bread and wine, there is an invocation of "the God and Master of all . . . hymned by the heavens and the heavens of heavens and all their powers, the sun and moon and all the choir of stars, earth, sea, and all that is in them." Or consider the language of the Epiphany services and its Blessing of the Waters, dating back to St. Sophronios in the seventh century, reliving the Blessing of the Jordan River and all the earth that took place during the Baptism of Christ, and customarily performed in bays and lakes and rivers: "Today the whole creation is watered by mystical streams . . . The sun sings Thy praises; the moon glorifies Thee; the stars supplicate before Thee; the light obeys Thee . . . At Thine Epiphany the whole creation sang Thy praises. . . . Today all nature is glad, things of heaven and things upon the earth."[26] And there is, too, the incomparable litany chanted during the morning Liturgy of Holy Saturday:

> Bless the Lord, sun and moon and stars of Heaven;
> Bless the Lord, light and darkness; nights and days;
> Bless the Lord, showers and dew and all winds. . .
> Bless the Lord, frosts and snows, lightnings and clouds;
> Bless the Lord, earth, mountains and hills, and all things growing in it

And the litany continues, including whales and fowls and animals, angels and prophets, and human beings, too. For Orthodox sensibilities, to worship God truly (*orthos doxa*) is to enter deeply into a cosmic community.

Sixth, the Eastern tradition teaches an appealing understanding of human psychology that can be found woven throughout the work of Dostoevsky. Inheriting and building on the distinction in ancient Greek philosophy between a higher and lower rationality—*nous* and *dianoia*—Byzantine psychology adds to it an understanding of the heart that is inspired by Jewish and Christian scripture, and corroborated by the ascetic practice of many traditions, including that of Sufism.[27] Given this analysis, the fallenness of our human condition consists in the malady whereby the *nous* (intellectual intuition, or more loosely, consciousness) has departed from its natural seat in the heart, and gotten tangled up in *dianoia*, discursive rationality—i.e., it has gotten lodged in the head.[28] But lacking the *nēpsis*—vigilance, mindfulness, or watchfulness—of *nous*, the heart has come to be ruled by desires, which in turn command discursive rationality, and along with it the *nous*, which nevertheless stubbornly believes in its own sovereignty. If this is the pathology of fallenness and sin, that of healing and redemption is simply for the *nous* to return to its natural seat in the heart, purifying the heart of the passions that have darkened it, and surrendering its delusions of autonomy as it is increasingly led by the uncreated light of divine energies.

This analysis offers us a key not only to understanding many of Dostoevsky's most important characters, but also his critique of Western Europe. The Underground Man, Stavrogin, Raskolnikov, Ivan, and a host of other figures are led into varying kinds of misery by their conscious identification with their thoughts—*logismoi* in the Byzantine lexicon. These thoughts make them proud and delude them into a sense of sovereignty and superiority, even as they become subtly aware that these *logismoi* are really being directed by the passions: anger, spite, envy, and so on. And to the extent that they are healed, there is a restoration of their conscious identity back to its natural home in the heart, and a resultant purifying of the heart: *metanoia* or repentance, and tears of contrition. And a reunion with nature. Western Europe, Dostoevsky believes, has valorized and enshrined the very principle of humanity's fall—the sovereignty of discursive rationality, which unknown to itself, is always the slave of desire—while denying the legitimacy of any higher knowledge or principle of illumination: refusing any direct contact of the soul with divinity.

This parallel between individual and society has powerful implications for understanding the human relationship to nature. In each of the figures just mentioned, and in many others, the sickness of the heart, the alienation of consciousness, and the rampant power of passion-led thoughts to disrupt individual life and social relations, all manifest themselves in a hostile, disrupted relation to the natural environment as well. Unable to see the divine energies in nature, and thus unable to see creation as *taxis*, or divine order, his characters rail against it, rebel against it, "spit on it" as the Underground Man puts it. They are at war with nature, and very much in need of falling to their knee in repentance, kissing the earth, and watering it with their tears. Yet it must be conceded that under the sway of *dianoia*—whether in the case of individual or society—the natural order can only be apprehended as the ironclad, oppressive "laws" of science, or else divine providence reductively transformed into utilitarian calculus of a sinister god, cold and manipulative, willing to sacrifice the happiness of an innocent child for the greater good of some greater number. Ivan thus rejects the notion of creation as a cosmic order, yet there is still something that moves his heart, giving him a reason to keep on living a little longer: the tender, sticky, little green leaves that appear in the spring stir in his heart a love of life.[29] And the reader feels that if Ivan's soul is to be saved—a pressing concern given his later conversation with the Devil, who shows that the proud sovereignty of his thinking has darker origins—it will be across this bridge of sticky little spring leaves.

There is, finally, a seventh principle of Byzantine thought and spirituality that importantly shapes Dostoevsky's thought. Going back to the third century desert mystic Evagrios of Pontos, and persisting in the Byzantine tradition to this day, is the concept of a threefold pathway of spiritual growth: It begins with the purification or purgation of the soul from desires and passions (*katharsis*), proceeds to the second step of illumination or noetic "seeing" (*theōria*), and concludes with ultimate divinization (*theōsis*) or uniting of the soul with the divine energies. Once the soul begins to be purified—and consequently, as Blake put it, once the doors of perception have been cleansed—we begin to see things as they are, begin to see all things in God as saturated with divine energies. Moreover, Evagrios and his successors emphasize that the initial stage of this illumination is *theōria physikē* or contemplation of nature—in its preliminary form, perhaps simply "seeing" nature, to use Annie Dillard's term—and in its fullest expression, seeing God in all things. And once again, we repeatedly encounter at least the first two steps of this threefold progression in Dostoevsky's writing, most prominently *in The Brothers Karamazov*—as Zosima's brother, then Zosima himself, then Zosima's mysterious stranger, and finally Alyosha, all begin to apprehend nature differently, once their hearts are purified, and all begin to see it as radiant with divine energies. Writes Nikitas Stithatos, in the *Philokalia*: "The soul's apprehension of the nature of things changes in accordance with its own inner state."[30] St. Anthony the Great also enjoins, "let us purify our mind [*nous*], for I believe that when the *nous* is completely pure and is in its natural state, it sees more clearly . . . since the Lord reveals things to it."[31] And conversely, to fail to see the divine energies in nature is to dishonor creation. After his illumination, Zosima's young brother apologizes to the very birds of the air: "Birds of God, joyful birds, you, too, must forgive me . . . There was so much of God's glory around me: birds, trees, meadows, sky, and I alone . . . dishonored everything, and *did not notice the beauty and glory of it all.*"[32]

The purification of the heart in both Dostoevsky and St. Isaac of Syria is not presented as something esoteric and arcane. Rather, its main presupposition is simply the acquisition of humility—a process undergone to one degree or another by virtually all of Dostoevsky's most sympathetic characters (Marya Timofeevna, Sonya Marmeladov, Prince Myshkin, Makar Dolgoruky, Elder Zosima, and Alyosha Karamazov) and visibly lacking in those displaying the pride that Dostoevsky feels comes from the Western subjection of the heart to calculative rationality (the Underground Man, Raskolnikov, Ivan Karamozov.) The humble person, in turn, attains a kind of rest that permits the rise of silent prayer and a certain kind of contemplation—called by St. Maximos *theōria physikē*, as we have seen in Chapter 8—which for St Isaac is always a gift of divine grace. The resulting, ecstatic contemplation of creation depicted by St. Isaac—what he describes as a noetic "vision of created things and the workings of providence among them"—seems like a spiritual portrait of the vision of the natural world described by Elder Zosima, while it also bears a striking resemblance to the "natural contemplation" we have seen in Muir, Hopkins, and many others:

> When someone reaches insights into creation on the path of his ascetic life [then] the person who has thus been illumined looks into all God's creation with the eye of the mind and sees there God's providence accompanying all things at all times [and] the moment he becomes aware of one of these mysteries, his heart is at once rendered serene with a kind of wonder . . . and it receives from grace the sweetness of the mysteries of God's wisdom and love. . . . At first you will find a joy which without cause overcomes your soul at times and seasons; and thereupon according to the degree of your purity *your eyes will be opened* to *see God's creative power and the beauty of created things*. And when the intellect will be guided by the wonder of this divine vision, then night and day will be as one to it because of its *awe at the glorious creations of God*.[33]

Thus, far from reverting to paganism, Dostoevsky is simply elaborating—with superb artistry—the authentic tradition of ancient Christian spirituality regarding creation, as preserved in its Byzantine and Russian transmission, that he had absorbed from the liturgies of the Church, learned from ascetical and monastic writers, and experienced first-hand during three pilgrimages to Optina Monastery, at the time the epicenter of Russian Orthodox spirituality.

But, finally, what about the kissing of the earth? Is this somehow pagan and idolatrous? Here, too, we can look to *Karamazov* for an answer, as well as to Dostoevsky's own journals and correspondence. We know from the latter that he saw this book as his culminating legacy to the world, as completing a project upon which he had long been at work, at least since his *Notes from the Underground* and its vision of The Crystal Palace, where we first find Dostoevsky's protest against Western rationality and technology that, parallel to Lynn White Jr., he saw as grounded in Latin Christianity, as is suggested rather explicitly in the parable of The Grand Inquisitor.

He called the subject of his critique "the reign of the man-God," and maintained that this modern modality represented the greatest threat to humanity, especially given its embodiment within the central project of Western rationality—that humanity should have reached a point at which the religion of the God-man and the consecration of nature (Christianity) was displaced and replaced by the religion of the man-God, humanity self-elevated to divine status, remaking nature after its own design, and indeed re-fashioning human nature according to its own calculations. These two alternatives, then, are in *Karamozov* set into conflict dramatically in the figures of two men: Ivan, who (apart from the sticky leaves) rejects the order of nature, showing himself to be strangely zealous to "vehemently deny God's creation, God's world and its significance"; and Zosima, who sees God everywhere within creation and whose joy in sharing this vision, even on the verge of death, seems irresistible.[34] In letters from May to September of 1879, Dostoevsky states repeatedly that he has put the strongest possible arguments in favor of atheism and the pointlessness of the natural order into the words of Ivan, with the hope of refuting them in a definitive manner, but *not* point by point. Not in the confident arguments of a discursive rationality that has gained sovereignty over *nous* and become severed from the heart, for he knew that it would lead inexorably to Ivan's conclusions. Not with some brilliant new version of the design argument. But aesthetically, spiritually, concretely, as embodied in the *life and person* of Zosima, and in whose *life* those readers who are able to see can clearly apprehend the refutation for themselves. Thus, like Plato, he saw that there are certain arguments that can only be refuted by life, never by more arguments, that is, by counter-arguments which could only lead to the skepticism that follows from impassible antinomies.

The most frequent locution naming the earth in *Karamazov* is the repeated phrase, "on earth." But why is this seemingly otiose qualification "on earth" so often appended to discourse over things both worldly and spiritual? Is there some reason that the planet be specified, as if there were other possibilities? The answer, I suggest, can be most simply found in the wording of the Lord's Prayer, which has immortalized the phrase: "on earth, *as it is in heaven.*" The invocation "on earth" thus reveals the earth as divinely ordered, affirms that heaven is the measure of earth. And hence, when in *Crime and Punishment* Sonya (whose given name is Sophia, Divine Wisdom) tells Raskolnikov (who indeed has appointed himself as a man-God) that he must kiss the earth and water it with his tears, she thereby tells him that his crime has offended grievously against the natural order by spilling the blood of his own kind, imposing an imperative to restore the lost harmony of heaven and earth, to assure that things "on earth" become now as they are in heaven. And for Alyosha, it is just as he emerges from his darkest despair that, "the silence of the earth seemed to merge with the silence of the heavens, the mystery of the earth touched the mystery of the stars"—i.e., when it *is* on earth as it is in heaven—that he throws himself onto the earth, embraces it, and, weeping, kisses it, just as his elder had recently enjoined him to do.[35] It is, then, not the earth as such, nor even less the earth as a pagan deity, but the earth *as it is in heaven*, the earth upon which "[God's] will be done on earth as it is in heaven"—the earth now revealed as infused with divine energies—that is honored, celebrated, and, indeed, venerated. It is the earth as a nexus of divine *logoi*, the earth seen in its iconicity—venerated just as Orthodox believers venerate icons at home and in Church, through bowing before them and kissing them, and loving not merely the icons but infinitely more what they reveal. And even if nature here is not yet fully restored to paradise, it still presents to those whose hearts have been purified a living anticipation of the New Jerusalem—an insight and sensibility that is inadequately grasped by Lynn White Jr. and contemporary disparagers of Christianity's relation to nature, and that offers a far more helpful vision of the relation of humanity and nature than its critics have supposed.

CHAPTER

10

Between Heaven and Earth

Did Christianity Cause Global Warming?

God empties himself
Into the earth like a cloud.

—*Annie Dillard*[1]

I

The evidence now seems overwhelming. Not only is the climate warming, but human activities have been an important cause of this process, perhaps even the principal cause. Meanwhile, global warming has replaced pollution, and even species depletion, as the preeminent symbol of environmental degradation. The ruthless pursuit and lavish expenditure of natural energies that have contributed to this problem stand as indictments of our shared complicity in environmental disruption, signs of modernity's "original sin" against nature.

Remarkably, nearly the same insight was being presented by Martin Heidegger as early as the nineteen fifties. Modern technology, he argued, constitutes an ontological assault upon nature, upon its integrity and soundness and its very capacity to manifest the holiness that the poet seeks to articulate—an assault that takes shape as the demand that the earth provide us with endless supplies of energy, and ultimately a disclosure of the being of the earth itself as energy reserve. In contrast to the traditional revealing of nature through *poiēsis*—the windmill and waterwheel, as much as the poetic verse—technology is an aggressive, provocative kind of revealing [*Herausfordern*] that imposes upon nature the "demand that it supply energy that can be extracted and stored as such."[2]

Even more than it is a logistical assault, with consequences yet to be foreseen, this is an ontological transformation, a change in the very being of nature itself: "Nature becomes a gigantic gasoline station, an energy source for modern technology and industry."[3]

Writing in the mid-fifties, Heidegger saw the greatest tangible threat of this metaphysical-technological energy-obsession to lie in controlling nuclear power, and this may yet turn out to be the case. But the consequences of planetary warming due to our extravagant combustion of fossil fuels now seems just as perilous—one more implication of our forcing nature to reveal itself as energy, and thus as inherently combustible and explosive, the all-consuming consumable. For Heidegger, this technological revealing of nature as energy supply had been long in the making, its origins reaching back to the time when nature or *physis* was first disclosed as *energeia* with Aristotle. But the ancient *energeia*, Heidegger hastens to add, is only a "distant" relative of the modern concept of energy, for it was still encountered within the Greek experience of *physis* as self-withholding self-emergence, as an interplay of presence and absence. It takes more than two millennia, Heidegger believes, for this ontological complicity of revealing and withdrawing to get reduced to the one-sided experience of presence in the modern notions of objectivity, resource, and energy supply. But perhaps the decisive moment, he maintains, comes during the Christian Middle Ages, when *energeia* gets thought by the Scholastics as *actualitas*, act and actuality. At the same time, nature gets understood as *ens creatum*, as created being, far removed from the experience of nature as self-emerging *physis*. Thus, to the extent that this one moment within what Heidegger understands to be a very complex history is truly pivotal, we could justifiably claim that Christianity, at least in this medieval Latin version, is in fact responsible for global warming, having prepared its metaphysical foundations.

But Heidegger was not alone in thinking about these things during the middle of the twentieth century. A half-planet away, in California, and almost certainly unbeknownst to Heidegger, the distinguished medieval historian Lynn White Jr. was also thinking a great deal about energy. And contrary to Heidegger, White found that the Western obsession with energy begins not in the twentieth century, nor even during the sixteenth and seventeenth century industrial revolution, but in the Latin Middle Ages and its "conscious and widespread programme [*sic*] designed to harness and direct the energies observable around us."[4] White's scholarly expertise, as discussed in earlier chapters, was in the history of medieval technology, and by examining the rise of one technology after another, he concluded that something remarkable happens between the tenth and fifteenth centuries: "a conscious and generalized lust for natural energy and its application to human purposes." Looking at a wide range of technologies—such as the compound crank, the belt-driven spinning wheel, and the artillery cannon—White sees a radical change in Western Europe's relation to nature, one that was to have planetary implications: "The expansion of Europe from 1492 onward was based in great measure upon Europe's high consumption of energy, with consequent productivity, economic weight, and military might."[5] And perhaps rather embarrassing for Heidegger, who somewhat nostalgically contrasts the "old windmill" with the hydroelectric generator, White shows that windmill technology first dates from this very period of medieval energy obsession, even seeing in it an epitome of the medieval appetite for harnessing the energies of nature. White's conclusion, however, is strikingly similar to Heidegger's, although he situates it not within the twentieth century, but in the apogee of this energy fixation in the Middle Ages: "They [in the thirteenth century] were coming to think of the cosmos as a vast reservoir of energies to be tapped and used according to human intentions. They were power-conscious to the point of fantasy."[6]

What specific feature of the Latin Middle Ages, then, resulted in this dramatic change in technologies? Surprisingly, White fails to identify anything peculiar to this time that might have brought about this energy obsession. In a later essay, reflecting upon the causes of these new technologies and the altered relation to nature that they entail, White departs from his exemplary historical scholarship and turns instead to freewheeling speculation. Ignoring important changes in Latin philosophy and theology, he maintains that what led medieval Latin civilization to become preoccupied with harnessing natural energies was just Christianity itself, pure and simple, just because its relation to nature is supposedly based squarely upon the Genesis injunction to "subdue and dominate" the earth. Even more remarkably, this stunningly simplistic conclusion has for some time been taken as definitive in the environmental literature, where it is cited routinely and uncritically.

What is so ingenuous about White's conclusion? A selection drawn from many possible objections should suffice. For example, both Judaism and Islam take the Genesis injunction as canonical, so why did these traditions not also give rise to modern technology and its energy-assault upon the earth? And if it is Christianity alone that is responsible, why does it take thirteen hundred years for the outcome to incubate? Moreover, why does this take place only in the Latin West, rather than in the Byzantine East, where as White readily acknowledges, a radically different relation to nature had always held sway, one based upon aesthetics rather than mechanics, sensing God as poet rather than engineer? What is special about the Latin West beyond its inheritance of Genesis I—which it shares with Judaism, Islam, and Eastern Christianity—that would account for this sudden craving for increased supplies of natural energy that has given us today our recent concerns with global climate change? When in his later essay White elaborates what he thinks are the attitudes and beliefs that led to these new technologies, he fails to provide anything specific to the Middle Ages that would be supported by his historical scholarship, but offers only the very familiar charges that by the fifties had already become something of a cliché: Religious otherworldliness, human exceptionalism, and repulsion at the earthly and bodily, all beliefs assumed as truisms among intellectual circles for whom figures such as Feuerbach, Darwin, and Nietzsche are canonical.

These charges, then, are generally familiar already to educated readers, for they simply recapitulate the tenets of Nietzsche's well-known attack upon Christianity, along with the humanistic reductionism initiated by Feuerbach and a biologism drawn from Darwin that been dominant for more than a century. But above all, it was Nietzsche, almost a hundred years before White, who most sharply charged Christianity with betraying the earth and the earthly in favor of another world, and thus too, with revulsion at the body. Yet for Nietzsche as much as for White, this putative otherworldliness is by no means peculiar to the Latin Middle Ages, nor even to Christianity proper, for he believes that it is rooted in the Christian appropriation of Platonism by early figures such as Augustine. This charge, in turn, is closely related to one made by his German contemporary, the Protestant theologian Adolf Harnack, who likewise claimed that Christianity had been subversively "Hellenized" during antiquity—an accusation robustly embraced by the early Heidegger, as discussed in Chapter 3. The problem, then, is really with Plato—and even if we overlook the fact that the first theological appropriation of Plato was undertaken not by a Christian thinker, but by a Jewish philosopher, Philo of Alexandria, not to mention the vast influence of Platonism upon Islamic philosophy, we would nevertheless be left with the question of why this purportedly Platonistic core within Christianity would have taken not *one* millennium but *two* to come to fruition in the medieval and modern quest for energy and power. Worse yet, it turns out that Nietzsche explicitly *celebrates* the quest for power or energy as the very essence of life and even being itself, and Heidegger sees Nietzsche not as the advocate of an ecological *Gelassenheit* that would tread gently upon the earth, but as laying the metaphysical foundation for the very technological assault upon nature itself. Heidegger, White, and Nietzsche all blame Christianity for betraying the earth, yet in ways that are so inconsistent with one another that the charge itself unravels into incoherence—a consequence, I suggest, of the fact that none of them were very interested in looking into the understandings of nature either in Latin Scholasticism or in the patristic thought of the early Christian period. So the question remains, what blame—if any, and of what sort—does Christianity incur for our present-day environmental crisis?

II

In 1637, Descartes proclaimed in his *Discourse on Method* that "we can have useful knowledge by which, cognizant of the *force and actions* of fire, water, air, the stars, the heavens and all the other bodies which surround us . . . we may be able to *apply* them in the same fashion to every use to which they are suited, and thus *make ourselves masters and possessors of nature*."[7] His admonition to master and possess nature through harnessing its energies—along with its metaphysical correlates of (a) *lifeless matter* whose very being is its extension in space, (b) a *distant deity*, and (c) an *imperialistic ego* with a keen sense of manifest destiny—all this stands at the mid-point of a certain history, concluding its Medieval phase and inaugurating its modern development. After Descartes comes a progressive inflation of subjectivity that culminates with Nietzsche, whose will to power assumes the status of world-creator, and whose Zarathustra proclaims that he cannot love what he has not created. But what must first happen to nature for human subjectivity to feel confident in assuming this prerogative of "making" itself "master and possessor" of it, a thought that then and even today (and perhaps, even especially, today) might well seem hyperbolic to the extent of madness? Mustn't the earth, if we are to draw upon Nietzsche's own image, first be cut loose from its sun, cast hurtling into darkness and disorder, before the New Prometheus can radiate upon it the illumining fire of its own subjectivity? It is, I will argue, a loosening of the commerce between heaven and earth; a waning of the interpenetration of visible with the invisible; a disenchantment of the earth, as Weber put it, that takes place in Latin Scholasticism—not as the advent of something entirely new, but as the culmination of a millennium of Western thought.

But long before Augustine and the Latin tradition that proceeds from him, the ontological foundations of early Christian thought were first worked out in the Christian East by thinkers steeped in the tradition of ancient Greek philosophy: in the philosophical schools of Alexandria, where Christian and pagan Neoplatonists for centuries maintained a lively exchange; in the wild deserts of Egypt and Palestine, where for the first time, with St. Anthony the Great, nature is experienced as a visible book, whose every word manifests the invisible Creator; and above all in the high plateau of Cappadocia, in central Anatolia, a decidedly rural area whose magical landscape of eroded rock formations resembles southern Utah, and whose lovely mountain valleys are equally evocative of the American West. The three great Cappadocian thinkers—St. Basil the Great; his brother, St. Gregory of Nyssa; and their friend, St. Gregory Nazianzus—worked out an ontology that differs dramatically from the later onto-theology of the West, so problematized by Heidegger, that sees God as the highest being, the metaphysical anchor of all other beings: the sun that Nietzsche intended with his analogy.[8] The Cappadocians saw God not as that kind of sun at all—as a Copernican, metaphysical axis to which the universe is tethered, and even less as a posited principle of intelligibility. The light they had in mind is rather the warm and gentle glow of gold leaf that embraces holy figures in the Byzantine icon, emanating from them as a divine luster, the uncreated light shining forth from deep within creation. These three thinkers saw all things as infused with divinity, as radiant with a holy integrity, that could truly be apprehended by those for whom, in Blake's words, "the doors of perception [had been] cleansed."

Drawing in radically new ways upon Aristotle's coinage, long assimilated into the Greek philosophical lexicon, the Cappadocians called this holy infusion into creation the divine *energeia*, the "energy" or "activity" of God displayed within things. Neither did this concept of divine energies—articulating God's immanence within the world—result from philosophical speculation or proceed from metaphysical motives, but from this new way in which it had become possible to experience nature and the soul. In contrast to the later *actualitas* of the Scholastics, and the conception of God as *actus*, the ultimate metaphysical grounding and explaining principle, the divine energies for the Cappadocians—and to this day in the tradition of Eastern Christianity, which has preserved their teachings—were incarnational rather than metaphysical, mystical rather than rational, meant to articulate the experience of seeing God in all things, an experience that they had already discovered through prayer and *askēsis*, the purification of the soul. Nietzsche was correct that asceticism was central to early Christianity, but wrong in judging its character: The early Christian ascetics sought vigorously, and indeed daringly, to cast off the artifice of the ancient urban centers—Antioch and Alexandria, and later Constantinople—in order to seek God in the wilderness through what they called *theōria physikē*, the contemplation of nature. The Russian philosopher Florensky in fact argues that the very "love of nature," the condition of what he calls "being in love with nature" that resurfaces in the West among Blake and the Romantics, derives its lineage from this very period of Christian asceticism. Nor is it merely coincidental that Basil himself is acknowledged by classicists to have been the first ancient author to express appreciation for the beauty of wilderness, in contrast to the pastoral landscapes of loved by poets such as Virgil. This sense of God in nature breaks forth in Basil's *Hexaemeron*, as he pauses to address the reader directly: "I want creation to penetrate you with so much admiration that everywhere, wherever you may be, the least plant may bring to you the clear remembrance of the Creator." The experience of nature upon which the Cappadocians drew in their concept of divine energies was sacramental, incarnational, and mystical—an orientation toward the *logoi* of every leaf and stone and living thing, the inner word through which they each, in an individual and unrepeatable manner, visibly articulate the Eternal *Logos*, the hymn and prayer that all creation *is*.

But was this experience and understanding not therefore at the same time pantheistic as well, making God a part of nature—or else panentheistic, making nature a part of God? Neither is the case, for to the divine energies (*energeiai*), these earliest patristic thinkers sharply contrasted the divine *ousia*, the divine essence or being, God as God is known by God. The divine being or *ousia* is radically mysterious, withdraws and conceals itself even from the highest angelic orders, remains always shrouded in eternal mystery. Rather than a metaphysics of presence, we find here an ontology of absence and withdrawal and concealment that lends itself to negative theology, and indeed historically gave birth to it, most splendidly in Dionysios the Areopagite, whose negations are intended—unlike those of Maimonides and Aquinas, and later, Derrida—not as conceptual tools, meant to leverage some analogy, or else dialectical strategies to sharpen some concept or other, but as vehicles for a mystical encounter with the divine energies themselves, in the very wake of this radical self-withdrawal of the divine essence.

Even in this brief exposition, it should be clear that this view is not by any means onto-theological, and this conclusion should have dramatic implications for our understanding of Western philosophy and theology, for it implies that neither the philosophical ideas of Athens nor the religious experience of Jerusalem necessarily result in the ways of metaphysical errancy, those *Holzwege* that Heidegger explored so persistently. Rather, the rich development of this ontology in the philosophy, theology, and spirituality of Eastern Christianity that can only be hinted at here—and which an unrelenting succession of conquests and captivities largely shielded protected from Latin influences—shows that there is indeed a path not taken by the West.

III

How, then, did the Western path diverge in this very different direction? How did the *"divine energies"* suffusing nature, epiphanies to be contemplated, the poetry of nature's creator, become *"natural energies"* there to be exploited for the sake of mastering nature itself, enabling us to make ourselves its ruler, if not its creator? In this chapter, a sketch will have to suffice for an answer.

We begin from a simple and retrospective hypothesis: The West could never find a way to affirm divine immanence without sacrificing divine transcendence, and was thereby forced to embrace the transcendent at the expense of the immanent, leaving nature in a state of abandonment. The classical distinction of divine essence or *ousia,* and divine energies or *energeiai*—one that would be richly elaborated by the great figures of Byzantine thought such as Maximos the Confessor, Saint Symeon the New Theologian, and Gregory Palamas—was simply never clearly grasped in the West, due in part to its diminishing ability to read the Greek texts, becoming chronic in the so-called Dark Ages, and forcing Western civilization to depend almost exclusively upon the prolific pages of Augustine's Latin texts for nearly a millennium. (A rare exception was Scotus Eriugena, whose anathematized thought reflects the work of St. Maximos the Confessor, whom he, in fact, translated.) Augustine, himself unread in Greek and unacquainted with the essence/energy distinction, was forced to reject the possibility of nature as theophany. Unable to understand how creation could manifest the divine *energies*, he had to conclude that the divine *essence*, being simple, could not be manifest within the spatially fragmented natural world at all, adding that this would compromise the divine simplicity. To encounter God, we must instead turn away from the visible world, like the prisoner in Plato's cave allegory turning his back on the cave wall (Conf. IV, 16) and look inside, within the *mens* or mind itself, where alone the divine light can be found. Even as he looks out upon the garden at Ostia—where together with his mother "for a fleeting instant" he "reached out in thought and touched the eternal Wisdom"—he notes that this ascent of thought demands that all that is "of earth, of water, and of air, should no longer speak to him" (Conf., IX, 10). Purification for Augustine becomes not the cleansing of the soul from passions, such as anger and lust, that would grant it the tranquility or *hesychia* needed for the contemplation of nature (*theōria physikē*), but rather the purification of the soul *from* everything bodily (i.e., from material creation itself). The will, for Augustine, must therefore will against nature, for it is the perversity of the will turning away from the creator and toward creation that is the source both of moral errancy and intellectual error.

By the time of Anselm, God is less a reality to be experienced than a hypothesis to be proved, and during the Middle Ages, "proving" what had never needed any proof now becomes something of a cottage industry, an endeavor that continues through the time of Descartes. By the time of Aquinas, nature is now self-subsistent, standing self-contained on its own, although supplemented and perfected by the supernatural, which nevertheless radically transcends it. The *analogia entis*, or analogy of being, still provides a connective link between the natural and the supernatural, but it is inferred rather than lived, thought rather than seen—operating like a fulcrum that in a limited sense allows thought, if not experience, to be catapulted into the invisible in search of conceptual treasure. *Energeia* is now *acutalitas*, the reality or actuality of nature, now understood not as epiphany, but as the actualized effect of a first cause who must, by that same fact, be entirely self-actualizing and fully self-actualized. Yes, there still remains analogy: the goodness of nature, its unity or coherence, its order and harmony, and above all its beauty all give us a proportional, analogical sense of the divine being, even if it is inferential rather than mystical or experiential. The *belief* that the beauty of nature expresses the divine glory is sustained, even if it has become problematic to *see* the latter in the former. But this last remaining link between creator and creation, between the visible and the invisible, was soon to be severed in two strokes. First, with Duns Scotus, who maintained the "univocity" of being, and went on to argue that since being is univocal, both God and nature "are" in the same way—that there is no higher mode of being or goodness or beauty in which nature would be rooted and inhere, and which it would in some way manifest. Nature begins to be "flattened" and lose its depth dimension, while gaining in self-sufficiency and self-subsistence. With Ockham, nature becomes even more radically autonomous, entirely self-referential, a purely extant realm of brute fact that lays the foundation for modern metaphysics. For Ockham, any inference or analogy from nature to God compromises the divine majesty and sovereignty, as does any suggestion that God chose to create this world, rather than some other, because it was good, beautiful, and so on. God created the world as it is simply because he willed it to be this way, and any attempt to arrive at insights about God through reflecting on creation is thus an impiety. Historians of science are correct that this lays the foundation for modern empiricism, for if nature is in no way expressive of anything higher, or more essential, then of course our only knowledge of it would be purely *a posteriori*. But it comes at the expense of reducing creation to a collection of facts and data, to which meaning must extraneously be assigned.

IV

It is important to note that only at this point do we encounter, in a fully developed form, the philosophical foundation for the ontotheology of Descartes and his metaphysics of nature as sheer extension, mute and lifeless matter ready to be plotted and calculated. That is, Western Christendom, for complex reasons, gradually departs from the classical, patristic ontology, and yet so powerful is this more originary understanding of God and nature—and its intertwining of immanence and transcendence—that almost to the end there remains a diminishing sense that nature is not just a product, *ens creatum*, but an epiphany and gift. Only in the final dissolution of the tradition, through the *via moderna* of Ockham, is this insight finally abandoned and the foundation laid for a metaphysic of nature as a field of forces that would support the pursuit of natural energies that, as White has shown, was actually taking place already.

Far from this metaphysic of nature representing a fulfillment of Christian thought and sensibility, as White maintains, it constitutes its inversion and antithesis. From *Genesis* to *Revelation*, creation is presented not as independent and autonomous, but as resting upon the providential power of the creator, shaped by the divine word, infused with the in-dwelling spirit hovering over waters. The evidence from patristic thought and the ascetic tradition, and their continuation in the Byzantine East, has been considered already. But the Western tradition itself gives subsequent internal evidence of how its view of nature has resulted not from a fulfillment of ancient Christianity, but from its dissolution and dethronement. The intellectual historian Michael Gillespie has shown that the willful, capricious, despotic God emerging from Ockham through Calvin—antithetical to the loving God, kenotic or self-emptying, of traditional Christianity—became the principal motive for modern atheism, itself sprung less from unbelief than from collective self-defense against a deity become a metaphysical monster.[9] For William Blake, this modern deity was Urizen, the impersonal, Deist God of Descartes and Newton and Locke, whose counterpart was an equally impersonal, mechanical, and mathematical nature, utterly impervious to warmth and love, repellent to the true "energies" of the imagination. For the early patristic thinkers and their ascetic counterparts, on the other hand, the knowledge of nature through *nous* and *theōria* was an event of heart and intuition, not of the discursive rationality that became increasingly dominant in the medieval West. And it was above all Blake who tried to retrieve this ancient way of knowing nature through what he called imagination, in contrast to the "empirical," calculable nature that displays not divinity but the cold laws of Urizen, a nature that Blake did not hesitate to call "*idolatrous*."

More recently, Jean-Luc Marion has engaged in a sustained reflection on the phenomenon of the idol, understanding it not as a particular kind of thing, but as the object of a certain kind of gaze or look—one that stands dazzled at the proximity of its object, stopped and fixated in such as way that it is reflected back upon itself.

The god of onto-theology, Marion concludes in sympathy with Heidegger, this highest, self-caused being, is truly an idol. But Marion seems less clear that onto-theological nature, whose prototype we find in late scholasticism, is equally idolatrous, for it is no longer encountered as *iconically* addressing the gaze with a visible revelation of what is inherently distant and invisible, but instead captivates and dazzles the gaze as an end in itself, even as it reflects back the look that apprehends it. For Marion, both icon and revelation as such are bound up with the gift, with what is given or *gratis*, and which theological reflection must ultimately regard as grace. To this, we may add the gift of divine grace as it is revealed in creation through the divine energies, everywhere manifest, but only to the "pure in heart" who the Beatitudes say "will see God." This graceful gift of divine energies stands in the starkest contrast to the seized-upon givenness of natural energies—whose emblem is not the lily of the field, but rather the gushing, overflowing oil derrick—under the idolatry of Western metaphysics and its bedazzled pursuit. Energy is thereby cherished and pursued and idolatrously venerated as enabling our own potentialities, as energizing us, as reflecting back to us the face of what we wish to be. This phenomenon of energy idolatry, then, leads us back to the same conclusion. Far from traditional Christianity being responsible for global warming, it is its very inversion and perversion into a secular metaphysic—the onto-theology that displaces and obscures traditional Christianity—that is to blame for these problems that are not just scientific and technological in character, but philosophical and ultimately spiritual. And it would seem clear that ultimately, their resolution must be also be found in this latter realm as well.

CHAPTER

11

Nature and Other Modern Idolatries

Kosmos, *Ktisis*, and Chaos in Environmental Philosophy

> How is this thing, this Newtonian phantasm . . . this Natural Religion, this impossible absurdity?
>
> —*William Blake, "Milton"*

> The Enlightenment fought with the opponent's weapons. In the place of a fantastic God, it put a fantastic deified Nature.
>
> —*Christos Yannaras,* Variations on the Song of Songs[1]

> We have to psychoanalyze science, purify it. . . . Its concept of Nature is often only an idol to which the scientist makes sacrifices, the reasons for which are due more to affective motivations than to scientific givens.
>
> —*Maurice Merleau-Ponty, "Nature"*[2]

A curious paradox with far-reaching implications. William Blake—poet, engraver, and great precursor of the Romantic view of nature—celebrated "the world of vegetation and generation," calling modern humanity both "to see a World in a Grain of Sand And a Heaven in a Wild Flower" and to take up arms against the "dark Satanic mills" of the Enlightenment worldview that suppressed our very ability to see.[3] How is it, then, that throughout Blake's writings, the very word "nature" itself accrues consistently negative connotations, even to the extent of being cast by him as debased and Satanic and inherently idolatrous?

The ancient Greeks, responding immediately and aesthetically, rather than indirectly and inferentially, had not referred to it as "nature" at all, but instead of *physis*, had said *kosmos*—visible adornment, manifest fittingness, displayed adornment. Neither did

Hellenistic Jews (in the Septuagint Bible) nor early Christians (in the New Testament), proceeding ontologically, call it *physis* or "nature," but instead had named it *ktisis*—creation, as manifestly ordered and overtly sustained by holy *Sophia* or divine *Logos*—or more often, they simply referred to "heaven and earth," demonstrating their conviction of the inseparability of these two realms, the celestial and the terrestrial.

So only in the seventeenth century did what had once been known as *kosmos* and as *ktisis* (rendered into English by such corresponding words as "world" and "universe" and "creation") enter our language as "nature": around 1662, according to the *Oxford English Dictionary*. Before that, the English word "nature" had been transitive, roughly synonymous with "essence," designating the nature or essence *of* something or other—a waning usage within a postmodern landscape that postulates things as *having* no nature, conveniently allowing us to make of them whatever we wish.[4] In the seventeenth century, then, nature begins to also designate the great aggregate of what is—an ontic region, an extant being, or to use the term made fashionable during this time by Descartes, an *object*.[5] As Heidegger maintains, the "domain called 'nature,' where you feel at home thinking in the way of the natural sciences, was first projected by Galileo and Newton."[6] And slowly, gradually, the more things become "nature," the more they lose their own natures: the *replacement* in usage corresponds to a *displacement* in ontology. This ontic aggregate, however, still exempts one mode of being from the totality, designating it (in English) by a word coined only some few decades earlier, i.e., "consciousness"—awareness or heedfulness, but now interiorized, as suggested by Locke's definition of consciousness as "perception of what passes in a man's own mind"—appearing in 1632, with "self-consciousness" gaining usage just a few years later.[7] This newly named and conceived dichotomy of nature and consciousness soon went on to become ensconced in Descartes's celebrated, reviled dualism *of res extensa* and *res cogitans* rendering nature (as extended) and consciousness (as self-grounding self-consciousness) binary terms. Yet new iterations were germinating quickly in this luxuriant, baroque milieu, for during these same decades of the seventeenth century Spinoza would maintain, contrary to Descartes, that consciousness and extension are merely differing attributes of the one substance—and that "nature" and "God" are in fact synonyms—allowing him to assert the identity of "*Deus, sive Natura*": God, that is to say, nature.

Thus, the triangular relation of God, consciousness, and nature that is constitutive of modernist rationality (and of metaphysics as onto-theology) is generated within a few decades of the mid-seventeenth century—complete with a newly minted lexicon. "Nature," rather than designating a mode of being, instead becomes reified and established as an ontic region unto itself, the reified realm of all that is not conscious. But meanwhile, nature so understood straightaway becomes identified with God. Almost as soon as it is circumscribed in this way, "nature" gets seen as inherently venerable, worthy of a kind of contemplative worship, and thus the object of what some would regard as an idolatry of the human intellect—which *itself*, in turn, becomes venerable for its ability to discern what looks very much like its own image in the mirror of nature. Thus, at the same time, we find the elevation of humanity to the status of a race of demigods—"the reign of the man-God," that Dostoevsky maintained was the core of modern nihilism, and the rule of "Lord Man" as Muir characterized this new self-understanding—that at the same time becomes evident within what would later be called "The Rise of Renaissance Humanism."[8] As noted by Gillespie, discussed in the previous chapter, in his more recent study of the roots of modern nihilism:

> What actually occurs in the course of modernity is not simply the erasure or disappearance of God but the transference of his attributes, essential powers, and capacities to other entities or realms of being. The so-called process of disenchantment is thus also a process of reenchantment in and through which *both man and nature* are infused with a number of attributes or powers previously ascribed to God. To put the matter more starkly, in the face of the long drawn out death of God, science can provide a coherent account of the whole only by making man or nature or both in some sense divine.[9]

But this modern reification and coordinate deification of what we now call "nature" is by no means confined to Spinoza, who simply anticipates and formally articulates a core modern sensibility that still remained largely covert, and whose complex unfolding is recounted brilliantly in Alexandre Koyre's masterpiece of intellectual history, *From Closed World to Infinite Universe*. Among other facets of this development, the Russian émigré Koyré shows how the cosmos—previously bounded and finite—comes to be understood and experienced as infinite, and thus as gradually appropriating for itself the attribute of infinity once singularly assigned to God. The "scientific revolution" was at the same time what Koyré terms "a very radical spiritual revolution," entailing no less than what he calls "the destruction of the Cosmos," i.e., the eviction of Western humanity from its perennial habitation in a world that was experienced as a harmonious, ordered, unified whole—a cosmos—and its displacement into a cold world of infinite, trackless matter that is now called "nature."[10] No less important player in this upheaval than Kepler expressed forcefully his own repugnance to what was taking place, as he became aware of this loss of the cosmos: "This very cogitation carries with it I don't know what secret, hidden horror; indeed one finds oneself wandering in this immensity, to which are denied limits and center and therefore also all determinate places."[11] But William Blake goes even farther, seeing here nothing less than a fall of humanity, its descent into idolatry. For him, the English inheritors of Descartes and Spinoza—notably Blake's infernal trinity of Bacon, Newton, and Locke—had led Albion (or England) into an idolatrous state: the worship of "the Goddess Nature"—the lifeless matter, to which the world had been reduced—and the concomitant fall into selfhood and selfishness: the reign of Satan, and the regime of pride and overweening rationality. Nature, locked up into itself and severed from both divine energies and human imagination, becomes an idol.

"Nature" is here, first of all, an artificial entity, the world deliberately stripped of the poetic—and if we are to use the word properly, bereft too of the symbolic, i.e., of that power that draws together (*syn*) into a whole (*holon*), and thus deprived of its beauty and coherence. Instead, this newly fabricated nature is coordinate solely to the *ratio*, to the calculative demands of the rational mind. Moreover, for Blake the fall into self and the idolatry of nature go together: Truncated nature and stunted intellect correspond to one another. Finally, with imagination reduced to the senses, and the world reduced to matter, there is at the same time a dissolution of both self and world. Without the unifying principle of what Blake called imagination—and what the ancients had once called *Nous*—the senses become not "doors of perception" upon the eternity of all things, but prisons of immanence: "Man has closed himself up, till he sees all things thro' narrow chinks of his cavern."[12] Henceforth, the world itself dissolves into Chaos, without an inner unifying principle—without form or meaning—a mere "Rocky fragment from eternity hurl'd."[13] This descent into "petrific, abominable chaos," which Blake understands as the wresting of "all things" from their inherence "in eternity," comes from the confused materialistic slumber and single vision of rationality become hegemonic.[14] Inner and outer, humanity and cosmos, are inherently linked here, coordinate terms of an unbreakable relation between them, which when dissolved renders both terms dissolute.

This relation between reason and nature was designated by a term that Blake thought amounted to an oxymoron: "natural religion" invoking the then (and still) fashionable nature venerated by deism. For Blake, deism (natural religion) represented the false and *artificial* religion of an hypostatized and idolized nature—self-contained and opaque and impervious to all transcendence—that was correlative to an equally contrived, remote and abstract God, related to "nature" only etiologically, as its one-time cause, and distant from any possible experience. This bifurcation of God and nature, he argued, radically diminished both terms: Nature as a vast idol becomes a cold realm of dead matter, the object of what he termed "Single vision and Newton's sleep," i.e., the result of seeing not "through" the eyes, but merely "with" them, as an optical object insensately sensing a perceptual object. Meanwhile, the deity becomes Urizen (i.e., "your reason"): the proud and vengeful demiurge, whose only relation to humanity is as "Nobodaddy" ("nobody's daddy," or perhaps, the daddy that nobody wants), the "Father of jealousy." In short, natural religion was for Blake an oxymoron because stripped of its interiority, of all inherence of divine energies, "nature" so-called was not in fact any more venerable than the distant deity who, driven away from all immanence, was merely an explanatory principle, known only inferentially. When held apart from one another, God and nature both become idols: an abstract, merely conceptual idol (God) and a material idol (nature).

But this bifurcation of the real into a lifeless idol called "nature," and a remote and abstract deity, even as it represents the main current of modernity, remains always incomplete. As Heidegger emphasized, always before (and always while) it becomes *natura* and *essentia*, *physis* is self-emergence, always already arising of its own accord, and thus overflowing the artificial canals of idolatry and onto-theology, engaging one person and then another. Thus, long after the seventeenth century, we still find, for example, Thoreau asserting that only the language of mythology was adequate to the richly textured experience of the woods and rivers as they presented themselves to him, and John Muir (along with a host of successors such as Annie Dillard and Wendell Berry) employing an overtly theological discourse to retrieve for his contemporaries an experience of wild places from which Western civilization (under the influence of deistic sensibilities now ensconced as metaphysical "naturalism") has long been disconnected. This interplay of divine immanence and divine transcendence was still at work, too, in many of the contemporaries and predecessors of Locke and Newton. Kepler felt strongly that the order and harmony of the cosmos expressed the deity, and saw the Holy Trinity symbolized in its proportions. And Copernicus so stresses the harmony of the cosmos that he believed was made more evident in his heliocentric universe, that Thomas Kuhn argues it was primarily upon aesthetic grounds that Copernicus's views were argued and, in fact, accepted.[15] With few exceptions, beginning with the Pre-Socratics, there has been a broad consensus in both philosophical and literary traditions that the harmonious unity of the world made manifest in a compelling manner something that both transcended the world and yet was everywhere immanent within it: *Nous, Logos, Sophia, Dēmiourgos, to Hen, ho Theos*. Indeed, even the skeptic Hume still felt obliged to acknowledge that the manifest unity of the cosmos, when regarded apart from the discord and fragmentation to be found in human communities, presented compelling evidence for a single author.[16] And of course, this is the great theme of creation myths everywhere: the emergence of an ordered, harmonious cosmos from the chaos that perennially threatens it.

The Greek word *kosmos* originally signifies order or structure (Homer refers to the *kosmos* of the horse the Achaians bring to Troy); then what it is that is itself ordered; and then the beauty and harmony *of* what is balanced and well-ordered, the evident charm of the fitting. Remi Brague puts it succinctly: "The term denotes order and beauty, even more specifically the beauty resulting from order."[17] It is important, then, that *kosmos* was eventually used to refer to the universe, for it suggests that rather than being seen as one set of characteristics among others, the interconnected order, harmony, and beauty of the universe was understood as demonstrating its most salient feature. Nor was this just an inference, a reasoned conclusion, but simply the way the world was apprehended. For example, in Book X of Plato's *Laws*, this ordered harmony and its reference to a transcendent source is stated to be evident first aesthetically; second by means of the universal belief that this is in fact true; and only third is it said that there are arguments as well, that the *kosmos* implies a creator, and that can be offered to those unpersuaded by the first two grounds. Nor is this perception of the *kosmos* as displaying this kind of lawfulness and order confined to the starry skies: in both Hesiod's *Works and Days* and Virgil's *Georgics*, we find frequent reminders of a cosmic order that needs to be discerned and respected, even if there is more emphasis upon the need to work against the tendency toward reversion into chaos and dissolution—as if it is the work of humanity to preserve the cosmos by laboring against the chaos out of which it was originally established.

The transparency of the universe toward a transcendent reality, however, is even stronger in the Jewish and Christian theistic traditions, for which "the heavens declare the glory of God." Creation, or *ktisis* (in both the Septuagint and in the New Testament), is always presented as displaying a transparent relation to its creator and sustainer, perhaps most impressively in God's answer to Job (not a set of reasons, but a stunning display of the sublime beauty of creation) and in the Sermon on the Mount (where the deep goodness and intricate interconnectedness of the natural world are employed to symbolize the human relation to God). Collating the Song of the Seraphim in Isaiah 6:3 ("the whole earth is full of his glory") with the final image in the Revelation of St. John 22:4 ("and they shall see his face [*prosōpon*]"), Biblical scholar Margaret Barker comments: "and we know from Isaiah that seeing the face of the LORD means recognizing the glory of the LORD in the whole earth."[18] Glory (*doxa*), in the Septuagint and New Testament, and then in the patristic and Byzantine usage, refers to something aesthetic, but this is an aesthetic of the first order, divine radiance streaming forth. In early uses (for example, Psalm 97:1–6 in the Septuagint) it is identified with the lightening that illumines the dark thundercloud from within and radiates forth into the world:

> Cloud and thick darkness are round about him; righteousness and judgment are the establishment of his throne. Fire shall go before him and blaze around his enemies. His lightnings appeared to the world; the earth saw and trembled. The mountains melted like wax before the face [*prosōpon*] of the Lord, before the face of the Lord of the whole earth. The heavens have [thus] declared his righteousness, and all the people have seen his glory [*doxa*].

Similar examples are found in Exodus, where Moses must hide from the radiant *doxa* of the Lord, and then in turn comes down from Sinai himself so radiant that people cannot bear to look upon him. But more commonly in the Psalms, the glory or *doxa* of God is seen radiating from and pulsating within the harmony and sublimity and grandeur of creation. Von Balthasar defines this *doxa*, or glory, as the "divinity of the Invisible, which radiates in the visibleness of Being in the world."[19] And of course it is surely what Gerard Manley Hopkins has in mind in his poem "God's Grandeur," where he writes that "the world is charged with the grandeur of God. It will flame out like shining from shook foil."

Thus, for both Athens and Jerusalem, the order and harmony and beauty of nature—as indications of a transcendent reality upon which they are founded, and which to those with eyes to see, can radiate out from within it—are by no means a proposition inferred, but rather a reality encountered. This is not onto-theological metaphysics, but lived experience. And if the word "metaphysics" is capable of recovery, we may want to say as well that it is *lived metaphysics.*

But it may be objected: Isn't this long-ubiquitous pre-modern orientation—which neither modernity nor post-modernity has been able to efface entirely—toward experiencing within the manifest unity and harmony of the natural world the immanent workings of a transcendent source—isn't this very experience of immanence itself the root of idolatry, the enticing endurance of paganism, the "natural" enemy of authentic monotheism? Isn't this primal orientation toward a sacred dimension sensed and discerned within our natural surroundings—a relation that is celebrated by David Abram, who nevertheless frames it misleadingly as "animistic"—inimical to monotheism's fundamental commitment to transcendence? Isn't it a movement of the soul outside the self, distracting it from interior realities of the soul where, according to some, we are supposed to look for God?[20] If we examine the analysis of idolatry in what can be taken as its classic refutation in *Against the Hellenes* by St. Athanasios the Great, we will discover, in fact, precisely the opposite: that here idolatry is seen to arise not from turning toward the cosmos, but in fact through turning away from it, back toward the self and its passions.[21]

Writing in fourth century Alexandria, where still-lively pagan sensibilities vied with Christian spirituality in the schools and streets, Athanasios argues that creation, *ktisis*, manifests to the uncorrupted soul, a compelling transparency to its creator. It is not just, as the Psalmist writes, the heavens that declare the glory of God, but rather *all* things natural speak of divinity to the pure in heart. On the one hand, our native experience of heaven and earth is one of transience and impermanence—a fragility of things, a tendency toward instability and disorder and dissolution that is evident whenever we consider them individually one by one. Yet our experience of the whole is radically different, for here we find instead balance and harmony, interrelation and interdependence: the same kind of features that were central to the ancient Greek experience of *kosmos*, and that led philosophers to articulate this unity by means of an inner organizing principle, whose transcendence of multiplicity and transience accounted for this unity of the *kosmos*. For Athanasios, this is as it should be, for he argues that this is the main purpose of the created order, to serve as a continual reminder of the creator—not a representation, but a *manifestation* of the divine goodness and beauty and unity. Nor is this an inference that we are somehow supposed to draw, but rather a discernible, perceptible identification that is comparable, Athanasios maintains, to the way one might recognize, i.e., might *see*, a particular sculpture as being the work of Phidias (35:1, 2). The Fall, therefore, is not a fall into nature, but rather a fall away from it and into ourselves. Athanasios argues that this occurs when we turn away from creation, which always draws us and entices us through its primordial transparency toward a transcendent creator, and turn instead toward what is our own (*idios*), toward what is closer to us—indeed, toward and into what *is* merely ourselves. But what is nearest of all, what is most our own, is the body, so that the fall into ourselves takes the form of a fall into the body and the merely sensual, the mutely sensual—a fall into what Blake would much later call experiencing the world with the senses, rather than through them, i.e., a fall into what he called "single vision." It is important, then, to note that Athanasios is by no means denigrating either the visible or the bodily as such. On the contrary, the visible is what is meant to serve as an ongoing gift of the divine presence, and the body is the means by which the *nous* (that is, our open and attentive awareness) can experience it and contemplate it and enter into a relation with the creator *by means of* the natural world.

Preferring what is nearer, staying close to shore and favoring our own—preferring the same to the other—we fall into ourselves; and this means we fall into the body as a prison or cave, as into a pit.

And here, alone within ourselves, we discover pleasure—and the desire for pleasure for its own sake—as a counterfeit of the soul's inherent, ecstatic urge toward transcendence. For as with Plato's account of *erōs*, and Husserl's theory of intentionality, Athansius sees motion as the inner principle of the soul. So that when the soul's motion is not directed toward the holiness manifest around it, and toward the transcendent creator that it announces, it moves instead to seek pleasure itself as its new (counterfeit) object, exchanging joy for pleasure, ecstasy for mere immanence, eternity for transience. Absorbed now in itself and its own desires, the soul nevertheless senses the need to find a terminus for its own desiring, an ultimate goal for the movement that is inherent to it. And thereby it creates for itself fictions, simulacra, idols that serve as loci of desire. And so it is, too, that idolatry is born. "For human beings," argues Athanasios, "having fallen into the unreasonableness of their passions and pleasures, and unable to see anything beyond pleasures and lusts of the flesh, inasmuch as they keep their mind [*nous*] in the midst of these irrational things, they imagined the divine principle to be in irrational things [themselves], and carved a number of gods to match the variety of their passions" (19:1). But since this turn away from God and that which is (*tô onta*) is a turn toward nothingness (*to mē on*) and dissolution, Athanasios argues that these fictions or simulacra become more numerous—more incoherent and at war with one another, as had become the Greek pantheon; more debased—with the worship of animals such as sheep and oxen and crocodiles, and even inanimate objects such as rivers and trees; and thus more demeaning to those who worship them. These idols eventually turn against the people who worship them, demanding they sacrifice their fellows to the fictions they have themselves concocted. The descent into idolatry—as a descent away from *what is* (*ktisis* or *kosmos*, along with the transcendent reality that constitutes it) into what is not—is thus a fall into dissolution and chaos.

I believe that the enduring viability of Athanasios's account of idolatry, and its importance for our understanding of the natural world, can be demonstrated through projecting three trajectories from this third century text to more recent thinkers: one focusing upon the positive, contemporary revaluation of idolatry in Feuerbach, Nietzsche, and Deleuze; a second briefly considering critiques of idolatry in Levinas and Marion in contrast to the classic account of Athanasios; and a third considering the work of Christos Yannaras, a contemporary Greek philosopher. Within the present chapter, the tracing of these strands will here have to be schematic and suggestive.

First, the positive appropriation of radical immanence and the attribution to it of the features of divinity. What became known as "naturalism" in twentieth century philosophy, the view that the real is nothing more than the object of the natural sciences, is a philosophy of immanence only by default. Seeing humanity as merely one animal species among others, it must regard religious aspirations as just one kind of animal behavior among others, a view that Feuerbach aptly called "vulgar materialism," for it underestimates matter itself. Feuerbach, in fact, plays a pivotal role in the creation of a philosophy of what we may call "radical immanence," i.e., a view that takes seriously the movement toward a sacred reality as a defining element of human existence, even as it seeks to embed this impulse within a sphere of pure immanence, and thus appropriate for itself the movement of idolatry as defined by Athanasios, expanding nature to include the characteristics previously attributed to the deity.

Accordingly, the concept of idolatry is important in Feuerbach's *Essence of Christianity*, which attempts to affirm what he sees as the truth of Christianity (and all religion) within a sphere of pure immanence. Idolatry here is taken in two senses: First, it is the projection of what is essentially human into an external sphere, i.e., seeing human nature as the feature of an external divinity. And in this sense, Feuerbach believes that we simply need to become more earnest, overt idolators, worshipping ourselves consciously and openly and joyously.[22] Thus, while he agrees with Athanasios that idolatry consists in projecting the human into a transcendent sphere, he disagrees with him over whether there is a more primordial truth that this projection displaces, and thus over whether or not idolatry is basically a good thing. In a second sense, however, he takes idolatry as simply the worship of nature *per se*, and here too he regards it positively, especially in comparison to Judaism and Christianity, which Feuerbach believes show disdain for the natural world. The nature-worshipper is, in contrast, a proto-scientist, exhibiting a somewhat childish form of the study of nature that becomes mature in modern natural science.[23] But since for Feuerbach human beings are, in fact, purely natural beings, the two senses of idolatry converge: the worship of nature culminates in the worship of nature's highest expression, the very worshipper himself. Metaphysically, too, the humanism of Feuerbach is founded solidly upon the concept of the idol. As argued by Marion:

> Subsidizing the absence of the divine, the idol makes the divine available, secures it, and in the end distorts it. . . . Everything is set up to allow a Feuerbachian reappropriation of the divine; since, like a mirror, the god reflects back to me my experience of the divine, why not reappropriate for myself what I attribute to the reflection of my own activity?[24]

Nietzsche, then, continues this positive revaluation and appropriation of idolatry, despite the casual reference to idolatry in the title of his final work, *The Twilight of the Idols*, which evokes both Bacon's use of the term "idol" to designate weak concepts given undeserved respect, and satirically evokes Wagner's *Twilight of the Gods*, *Die Götterdämmerung.* For like Feuerbach, he believes that all gods are idols—which for him does not imply that we ought not to create gods, but only that we should create newer, better ones, and he complains that in two thousand years we have not created a single new god. What is problematic for Nietzsche, then, is not the manufacture of deities and thus idolatry as such, but rather botching the job—not the worship of fictions, but the feeble imagination employed in making them. Thus, as in Feuerbach, we need to be bolder, better idolaters, and fabricate worthier deities than those that are threadbare and outworn. The figure of "Dionysios" is Nietzsche's own creative idol of choice, while his overall strategy for sanctifying radical immanence in general is the doctrine of the eternal recurrence of the same, stamping all transience with the holy "seal" of what Zarathustra calls "eternity, deep, deep eternity"—a stamp that is meant to legitimize the veneration of everything earthly and transitory, even what is petty and small and fleeting: that which Athanasios (like Nietzsche) saw as tending of itself toward dissolution. Moreover, Nietzsche adds a further element to the reappraisal of idolatry, through his positive evaluation of what he calls "chaos," as a prerequisite for genuine creativity within the horizon of radical immanence, of being "true to the earth, "and thus properly venerating the idol of "nature." For the earth, too, is itself a product of the creative energies of the *Übermensch* and his revaluation of immanence.

But the metaphysics of radical immanence is carried to an ultimate, and rather prodigious, conclusion by Gilles Deleuze, sometimes in collaboration with psychoanalyst Felix Guattari. Deleuze's claims in this regard are remarkably consonant with the fourth century analysis of idolatry by Athanasios. Yet while the respective taxonomies are often the same, their appraisals differ radically. What for the ancient sage is self-evidently repellent, the postmodern luminary regards as salutary. The fall-into-self, diagnosed by Athanasios as the very origin of idolatry, is positively embraced by Deleuze, who regards what he calls (after Artaud) the "body without organs" to be exemplary, arguing that it constitutes the best way to make sure that the self is inaccessible to God—and above all to the "judgment" of a God who is experienced not as a blessing, but as Blake's menacing Nobodaddy. Immersion into a multiplicity of desires is positively re-valued by Deleuze—perhaps for the first time in the history of philosophical thought—who sees human beings, along with other natural configurations, as desiring-machines, for whom the movement of desire is intrinsically self-legitimating, as lips, genitalia, breasts, and so on mechanically hook up with one another in a faceless saturnalia of the senses. That this absorption in sensation and desire is faceless and depersonalizing is not seen as alarming, but rather is valorized by Deleuze—while the face itself (which he believes holds an essential connection with the face of Christ, i.e., of a transcendent God become immanent) is seen as menacing and oppressive. Likewise, he finds the natural landscape constrictive and oppressive, seeing it too as entailing a kind of faciality. And we may recall here Athanasios's suggestion that we can recognize the creator from the world around us in the same way one recognizes the sculptor Phidias in encountering his work—solid grounds, indeed, for someone with beliefs such as those embraced by Deleuze to turn away from the natural world. Better, perhaps, to cling to cafés where the soothing sight of many heads, seen mostly from the back and thus as faceless, would constitute a more benign environment.

But the inverted mirror image—where left becomes right, and vice versa—continues. Like Athanasios, he regards the relation to a transcendent deity as the ultimate source of the unity of the person, but for Deleuze this is the very reason we need to *turn away* from this relation. Beyond this, Deleuze elevates the simulacrum over the image: While the image evokes what is other, the simulacrum substitutes for it, displaces it, replaces it. Finally, the notion of a primordially ordered cosmos is rejected altogether, and chaos itself is seen as venerable; indeed, he reduces the former to the latter in his hybrid term "chaosmos," i.e., the cosmos seen as itself not being order, but rather chaos. The objectification, and deification, of "nature" and immanence—a "chaosmic" world with no apertures of transcendence, closed in upon itself like a macrocosmic "body without organs"—has reached its completion and fulfillment (or seen differently, its final *reductio ad absurdum*) with Deleuze.

In the sharpest contrast, the immersion in the same, the refusal of transcendence, and the embracing of idolatry are seen by both Levinas and Marion as what is most to be avoided, among the most insidious outcomes of modernity. Levinas's writings develop through many variations the themes of alterity or otherness, and of the ethical demand that we not reduce the alterity of the other to the ipseity of the same, as he believes takes place simultaneously in the philosophy of being, identity, and *totality* exemplified by Hegel and Heidegger, the politics of totalitarianism, and the state of war. (It is significant, looking back, that Deleuze assigns a positive valence to what he calls "war-machines." Significant, but hardly surprising, since as David Bentley Hart observes, "desire, for Deleuze, is power, and delight is always somehow an aggression."[25]) Only the metaphysics of *infinity* can preserve the face of the other, while God must remain radically transcendent, infinitely distant. But this would mean that any sense of divine immanence within the world is suspect, presumptively illegitimate and necessarily idolatrous, and Levinas draws this conclusion boldly, arguing that Judaism has rightfully "decharmed the world," and shown that any sense of "the Sacred" as experienced within the world is "the essence of idolatry" and "a form of violence."[26]

Thus, in order to safeguard the otherness of God and divine transcendence, all immanence is here stigmatized by Levinas, leaving us with a one-sided transcendence corresponding to Deleuze's one-dimensional immanence. Building on the work of Levinas, Jean-Luc Marion argues for the important distinction between the idol (reflecting, as Athanasios had argued, the gaze of the idolator) and the icon, which counter to Levinas's view, would allow transcendence to enter the world without being subsumed by it. But through his emphasis on distance, Marion seems to confine the iconic manifestation to the face alone, thus closing off nature itself as a medium of theophany, in effect relegating it to the sphere of the idolatrous. Moreover, neither Levinas nor Marion see idolatry as being a corruption of something more original, away from which we have fallen, but rather both seem to regard idolatry as itself the default state, against which we must be ever vigilant. Thus, both (to different degrees) leave us with a cosmos that is perhaps just as one-dimensional, and quite as bereft of transcendence, as the world of Deleuze and Nietzsche and Feuerbach, and prior to them the deistic philosophers against whom Blake took up arms. The horns of our postmodern dilemma would thus be, on the one hand, an objectified nature, itself become deified and idolatrous (best exemplified by Deleuze), and a nature bereft of holiness, cast unto itself, in deference to a distant deity (best exemplified by Levinas). Is there a way past these two alternatives?

Onto-theology loses neither its theological character nor its ontic orientation when nature is substituted for God as the normative being. But what if neither God nor nature were beings at all—if the concept of beingness or essence (*ousia*) was itself a radical distortion of the mode of being of both God and nature? This question is pursued powerfully and originally by Christos Yannaras, a contemporary Greek philosopher whose work (long accessible in German) is only now being translated into English, especially his masterwork, *Person and Eros*. Working both from Heidegger's critique of metaphysics (which he argues remains incomplete) and from Greek patristic thinkers of the first millennium, who he sees as completing the philosophical project of ancient Greek philosophy, Yannaras presents a compelling rethinking of the Heideggerian *Seinsgeschichte*, arguing that Heidegger—whom he calls "perhaps the last 'essence mystic' in the West"—in his understanding of ontological difference, ultimately substitutes an *apophatic* essentialism for the discursive, *kataphatic* essentialism of metaphysics and onto-theology.[27] Difference or otherness, argues Yannaras, must instead be understood as arising more radically from within the experience of referentiality or relationality—within an apophaticism not of essence, but of relation—whose locus is necessarily that of the person.

But it is crucial here to understand "person" not as it is distorted by the Latin term *persona*—the mask that conceals true essence and authentic *substantia*—but rather through the original Greek *prosōpon*, initially face or countenance, derived from compounding the preposition *pros* ("toward) and the noun *ops* (eye, as in "optic," or face), and thus signifying: to "have my face turned toward someone or something." That is, person as *pros-ōpon* entails "immediate reference, a relationship."[28] The radical character of Greek patristic thought, then, can be approached through the insight that what truly stands on its own, *hypo-stasis*, is not *physis* or *ousia*—essence or nature or substance—but rather relationality and referentiality; not what stands on its own, but that which is inherently relational. In contrast, Latin thinkers from Augustine through the Scholastics understood *prosōpon* as *persona*, originally the mask put on by actors, and thus as exterior and subservient to the true, underlying reality or essence; hence, the inevitable regression back to essentialism in Scholastic theology, i.e., to getting behind and beyond the mask or exterior. Beyond this, the "towards something" (*pros ti*) of the *prosōpon* radicalizes Heidegger's ecstatic character of *Dasein*, extending the *ekstasis* into a mode of disclosure that constitutes itself as a response to an invitation (*klēsis*)—i.e., it must be understood as *erōs*. And correspondingly, reciprocally, the phenomenon itself "*is disclosed (phainetai)* as that which it *is* only in the fact of the relation that reveals the otherness of the person."[29]

What does it mean to think nature or *physis*, then, within the context of primordial relationality? For Yannaras, it means that *physis* as such must be understood through those modes of existence in which nature or essence is self-transcending, entailing not just freedom, but the emergence of otherness itself—i.e., of *to idiazon*, of that which makes distinctive from the essence as such. But if this is the case, it could only be through what transcends the natural—both the transcendence of God, and that of humanity, and of the relationality within which reciprocal, if non-symmetrical, transcendence is exercised—that nature itself is revealed. And this, for Yannaras, as for Heidegger, entails a poetic revealing, a shining-forth in beauty, and therefore above all the disclosing of the personal reality revealed within the natural, just as much as the painter and person van Gogh is revealed and recognizable within each of his works—as Athanasios had claimed with regard to the ancient sculptor Phidias. Moreover, Yannaras argues that our ability to respond to this personal reality disclosed through and *within* the natural is a correlate of our own transcendence of nature, which unlike Kantian morality entails not the suppression of the natural but its elevation within an erotic movement.

Thus, the opacity of reified, idolatrous "nature" would be precisely what renders this disclosing marginal, if not extinguished it altogether. The idolatry of nature necessarily obscures and disfigures precisely what it believes itself to be elevating: nature itself. Its face hidden, it appears to be a mere substance, whose presence does not encounter and address us with any meaning, but is "just there" or as Heidegger puts it, *vorhanden*. The idolatry of modern metaphysics, Yannaras maintains, thus follows from the failure of the Latin West to understand the primacy of the sphere of *prosōpon* and relationality and ultimately *erōs*, and its reversion to the metaphysics of essence and substance and distanced (objectifying) knowledge.

Can the natural world—the cosmos as we encounter it—present to us something like a face, and thereby disclose a depth, a warmth, a light both from within and from beyond itself, not as an onto-theological postulate, but as a lived encounter? This very cursory and condensed presentation of Yannaras's rich and nuanced thought will likely give only a suggestion of what this might well serve as a solution to the problem of modern, idolatrous nature—a solution that avoids the Scylla of utter immanence and the Charybdis of pure transcendence. But what may nevertheless be clear, I believe, is that the re-solution must entail neither radical immanence nor pure transcendence, but rather an *immanent transcendence*, and it is for this reason that concepts such as *erōs*, transcendence, relationality, referentiality, *kenōsis* or self-emptying, and indeed intentionality itself all provide important indicators of how the former is to be understood. Moreover, I believe that it is possible to find within the roots of Western thought itself an expression of this immanent transcendence that counters the subsequent Western attachment to substance and essence, to reification and objectification, that obscures what the perennial wisdom of humanity has always grasped, however dimly, i.e., to find there as well (there where the danger is greatest, to paraphrase Hölderlin) in perhaps a definitive manner, the salvific and healing element for which we, and the natural itself, "groan and travail" (to paraphrase St. Paul) in a technological, postmodern age.

CHAPTER

12

Traces of Divine Fragrance, Droplets of Divine Love

The Beauty of Visible Creation in Byzantine Thought and Spirituality

> And all of these refer to some slight *trace of the divine fragrance*, which the whole creation, after the manner of a jar for ointments, imitates within itself by the wonders that are seen within it.
>
> —*St. Gregory of Nyssa*

> All things around us are *droplets of the love of God*—both things animate and inanimate, the plants and the animals, the birds and the mountains, the sea and the sunset and the starry sky. They are little loves through which we attain to the great Love that is Christ . . ."
>
> —*Elder Porphyrios,* Wounded by Love, *218*

I

In discussing beauty, which is extraordinary, I want to begin from everyday, ordinary experience, to suggest that ordinariness itself is a constraint we heedlessly impose upon the extraordinary. I want to begin with the small owl unexpectedly encountered, bathing in a pool of water after a rain, whose beauty illumines the remainder of the evening with a certain charm, a spiritual fragrance of enchantment—or with the dusty, late afternoon sky glimpsed momentarily along a country road long ago, whose muted, translucent hues are beautiful in some subtle, but deeply satisfying way, giving rise to a distinct sense that this wonderful but inconspicuous beauty, which seemed to come from somewhere else, was connected to an unending source of goodness that could sustain the soul forever.

These ordinary glimpses of beauty that we are granted from time to time are unsought and unexpected: unlike the beauty of the

Louvre or the Bernese Alps, which we intentionally, expectantly seek out. Their ultimate charm, I suspect, lies in their tacit suggestion that our *every* experience of creation has the potential to be extraordinary, beautiful, sublime, if only our souls were prepared for it, if we had the eyes to see.

The ancient Greeks had such eyes. Perhaps to a fault. They saw the extraordinary in the ordinary: In owls and lightning flashes, in the ecstatic exhilaration of the hunt and in the quiet goodness of tender shoots rising up from the earth, they saw the invisible rising up into the visible; they saw the beautiful. And because the beautiful moves and attracts, is *attract-ive,* they were drawn to it so powerfully that they worshiped it everywhere, for they found it all around them. At the dawn of Greek philosophical thought, Thales of Miletus gave expression to this sensibility, according to Aristotle (*De An.* 411 a7–8) saying that all things are full of gods.

It is this dispersed, fragmented, incoherent worship—and view of nature, upon which it was based—that the early Christians denounced as idolatry—in St. Paul's terms, worshipping and serving the creature rather than the Creator (Rom 1:35). But for critics from Nietzsche to Lynn White Jr., this very denunciation was itself culpable: It was Christianity, they maintain, that subverted this laudable veneration of nature, turned its back on the beauty of the visible, seeking instead to escape into another, super-natural and otherworldly, goodness and beauty, and thereby inaugurating a devaluation of nature that would eventually culminate in the environmental crisis of today.

II

The philosophical response to environmental concerns first took shape in the early 1970s, as the search for an environmental ethic. Responses in Christian theology came a few years earlier, and they too were heavily oriented toward moral concerns. But it has by now become evident to many that an exclusively ethical orientation—a narrow focus on concepts of rights and duties, as well as the more theologically grounded notions of responsibility and stewardship—is limited, and perhaps also distorts something important about nature that escapes the moral outlook. The moral approach is limited, first of all, because it inevitably leads to conflicts between competing moral demands—for example, concerns for jobs and economic development versus spccies preservation versus green space and recreation, and so on. But it is also limited because it seems to grant nature no greater status than that of a resource whose value lies in satisfying human needs, whether or not this commodity status is understood to be mediated through a stewardship relation to God. But something is missing when we see the created order only in its usefulness—a sense of its integrity, of its sanctity, of its beauty. Perhaps, then, an aesthetic approach to thinking about the natural environment has something vitally important to add. Moreover, in contrast to a strained application of the modernist lexicon of rights and duties and obligations to environmental problems that they were never meant to address, our native intuitions effortlessly and spontaneously support a respect for nature that is based on admiration for its beauty. Wanton destruction of a natural landscape (or even a single flower) offends our aesthetic sensibilities more than it strikes us as immoral, as would (for example) the ill-treatment of a person. But against the grain of modern philosophy, this aesthetic imperative would need to go beyond such subjective notions as "taste" and "enjoyment," instead deriving its normative power from the ontological standing of beauty itself—from the kind of autonomous, and indeed commanding, being or mode of existence that is specific to beauty in general, and more specifically to natural beauty.

Where to look for an aesthetic that would understand the beauty of nature ontologically, as entailing correlative, practical imperatives? Perhaps in some neo-paganism, as followers of philosophers such as Nietzsche, and to some extent Heidegger, have rather ingenuously suggested? Or do we have no other alternative than to look to other cultural traditions—to Taoism or Buddhism, for example? Yet no current of thought, East or West, has displayed a greater proclivity for the beauty of the material world—and more insistently seen it as demanding a practical response—than has Orthodox Christianity, which has always possessed a deep, sacramental orientation not just toward the bread and wine of the Eucharist, the water of baptism, and the oil of chrism, but toward the material world as a whole. It appears in the beauty of Orthodox temples and iconography, of its chant and liturgy. As Dionysios the Areopagite puts it, the sensuous character of the Liturgy—the fragrance of the incense, the warm luminosity of the candles, the splendors of the iconography, the sonorous loveliness of the chanting, and at its culmination the very taste of the Eucharistic bread and wine on the tongue—all are meant to strike the senses as noetic emanations, as gifts of immaterial light, and ultimately invitations to an ontological participation in the metaphysical locus of Beauty, in Christ Himself, with whom the believer becomes one in the Eucharist (*Celestial Hierarchy* 121c–124a). Moreover, this iconicity extends to the entire cosmos. "The great, all-shining, ever-lighting sun," Dionysios proposes as an example, "is the luminous icon of the Divine Goodness" (*Divine Names* 697c). "Truly," he affirms, "the visible is the manifest icon of the invisible" (*Epistles* 10, 1117B).

The Byzantine predilection for beauty—its *philokalia*, to use a term in currency since the time of St. Basil the Great—its inclination toward the beautiful, extends far beyond temple and monastery and icon corner into an overall orientation toward nature as a whole, a sensibility it has inherited from the Patristic experience of nature—an emended, augmented, but nevertheless continuous legacy from the sensibilities of the ancient Greeks. For contrary to the romanticized revisionism of Lynn White's claim—based on his reading of the Latin West—that Christianity disrupted a happy pagan love for nature, in fact it was not pagan antiquity at all, but early Christianity that first cultivated the very love for nature, the "feel for nature," the appreciation of the beauty of nature for its own sake, that not only inspired the romantic movement of the nineteenth century, but continues to animate environmentalism today.

This point was argued persuasively by Fr. Pavel Florensky in his *Pillar and Ground of Truth*, perhaps the great neglected work of twentieth-century philosophy, which has only recently been translated into English. Countering Nietzsche's valorization of the Greek enchantment with surfaces and appearances, Florensky argued that paganism hugged the surfaces, dwelt upon appearances, due mostly to its fear of the demonic realities felt to be lurking beneath and behind the beautiful forms—namely, the gods whom they sensed everywhere, and everywhere feared, and whom they thus strove to "magically control" through sacrifice and ritual.[1] Thales's maxim that all things are full of gods, given Florensky's rendering, would need to be understood less as an expression of misty-eyed veneration, and more like a battlefield warning that one step in the wrong direction might trigger a landmine. We would be led to reappraise pagan nature-aestheticism with the keen eye of the Septuagint Psalms, which assert that "all the gods of the nations are demons [*daimonia*], while the Lord made the heavens" (Ps. 95). In contrast, Florensky maintains that "only Christianity has given birth to an unprecedented being-in-love with creation . . . If we take the 'sense of nature' to mean a relation to creation itself, not to its [superficial] forms, if we see in this sense more than an external, subjectively aesthetic admiration of 'the beauties of nature,' this sense is then wholly Christian and utterly inconceivable outside of Christianity."[2]

How was this changed relation to creation accomplished, this transition from fear to wonder? Florensky, himself both a scientist and a historian of science, argues that only with the rise of ancient Christianity does nature have an inner reality of its own—have its own intimate relationship to the God who has created it—and thus possess an ontological weight, a reality proper to it, as distinct from serving merely as a mask for some shrouded deity—prerequisites for both the love of nature, and later the science of nature. "This relation to nature," he maintains, "became conceivable only when people saw in creation not merely a demonic shell, not some emanation of Divinity, not some illusory appearance of God, like a rainbow in a spray of water, but an independent, autonomous, and responsible creation of God, beloved of God and capable of responding to His love."[3] Only now does nature become fully real, and truly lovable, and ultimately intelligible.

But what about the asceticism—world-denying and earth-despising—that Nietzsche and his followers have excoriated? Florensky helpfully distinguishes between two asceticisms. On the one hand, authentic Christian asceticism is in love with God's creation, and everywhere encounters the uncorrupted beauty of nature, looking within the outward forms that have become distorted in a fallen world of corruption and death. In contrast, however, are those other asceticisms—ranging from Vedanta to the world-weariness of Tolstoy and the Russian intelligentsia—that reject the world altogether, not just its distortions. These truly do represent an "escape" from a world found disgusting and repellent, a repudiation that Florensky insists is a blaspheming of God's creation.

But beyond this, it is precisely in the most perfected forms of Christian asceticism that we find the greatest love for the beauty of creation. Why would this be the case? Above all, because the ability to know and the ability to love both entail spiritual prerequisites that are perfected in monasticism. Writes Nikitas Stithatos, in the *Philokalia*: "The soul's apprehension of the nature of things changes in accordance with its own inner state."[4] *Changes (of perception) in accordance with its own inner state.* St. Anthony the Great also enjoins, "Let us purify our mind [*nous*], for I believe that when the *nous* is completely pure and is in its natural state, it sees more clearly and further . . . since the Lord reveals things to it."[5] And more recently, Elder Porphyrios: "[Not] everyone here sees the light of truth with the same clarity. Each person sees according to the state of his soul. . . ."[6] Even a visitor to Mount Athos, after spending time with Elder Paisios, reports that the world of creation around him now presents itself quite differently: "When we stood up to leave, we felt nature taking on a different appearance around us. We sensed everything spiritually. It seemed like even the plants might speak to us . . ."[7] It is the purified, monastic consciousness, always exemplary for ancient Christianity, that sees most deeply the goodness and beauty of nature, and hence is most able to love it. "Blessing the universe, the ascetic everywhere and always sees in things God's signs and letters," says Florensky. Thus, it is neither Galileo nor Leonardo who first regards nature as a text, but St. Anthony: "My book is this created nature. It is always with me, and when I wish I can read in it the words of God."[8] "The higher the Christian ascetic ascends on his path to the heavenly land," writes Florensky, "the brighter his inner eye shines, the deeper the Holy Spirit descends into his heart—the more clearly then will he see the inner, absolutely valuable core of creation . . . It is precisely among the charismatics and ascetics that we find the most striking examples of a feeling that I can only call the *being-in-love with creation*."[9]

Florensky's claim—that with Christianity, and especially Christian asceticism, a new sensitivity for the beauty of creation emerges into human awareness—is borne out strikingly in protocols from the great ascetics of the Orthodox East and their successors, up to the present day. It is, for example, in a letter of St. Basil the Great to his friend St. Gregory of Nazianzus that we find what is likely the first example in antiquity of praise for a natural beauty that is not pastoral, but wild and uncultivated.[10] Savoring the beauty of his Cappadocian hermitage overlooking the River Isis, he describes "a lofty mountain covered with thick forest," "cool and transparent" mountain streams, precipices and ravines, and the great river itself with a swifter current than he has ever seen before, cascading dramatically onto rocks and forming deep whirlpools—a far reach from the docile, domesticated beauty of Hesiod or Virgil.[11] And in his *Hexaemeron*, Basil pauses to reflect on how the "beauty and grandeur" of creation—"earth, air, sky, water, day, night, all visible things"—is a "training ground" for the soul to "learn to know God, since by the sight of [these] visible and sensible things the *nous* is led, as by a hand, to the contemplation of invisible things."[12] Later, he interrupts his narrative once to again address his reader directly: "I want creation to penetrate you with so much admiration that everywhere, wherever you may be, the least plant may bring to you the clear remembrance of the Creator."[13] Or we may listen to his brother, St. Gregory of Nyssa, extolling the emergence of springtime from the harsh winters of the Cappadocian plateau, while evoking an image of the Resurrection: "[Springtime here does not] shine forth in its radiant beauty all at once, but as preludes of spring [come] the sunbeam gently warming the earth's frozen surface, and the bud half hidden beneath the clod, and breezes blowing over the earth."[14]

Nor has this love of nature's beauty cooled or diminished in the Orthodox asceticism of modern times. "The holy Athonite Fathers," proclaim the Representatives of the Twenty Monasteries, "have testified to the way in which their communion with supernatural spiritual states led them to feel a special affection and concern for their natural surroundings and to experience a spiritual sense of harmony with the whole of nature."[15] Listen to the words of Elder Ephraim, recalling those of St. Gregory: "Now in springtime, when nature is wearing its most beautiful apparel, one feels inexpressible joy when this natural beauty is accompanied by a sublime spiritual state. Truly, our holy God has made all things in wisdom! *The soul cannot get enough of beholding the beauty of nature.*"[16] The Paschal theme is also echoed poignantly by Elder Porphyrios: "I looked at the clear, blue sky, at the sea which stretched out endlessly, at the trees, the birds, the butterflies and all the beauties of nature, and I shouted full of enthusiasm: Christ is Risen!"[17] And who could match the tender aesthetic of the heart expressed by Elder Joseph the Hesychast?

> Listen to the rough crags, those mystical and silent theologians, which expound deep thoughts and guide the heart and *nous* towards the Creator. After spring it is beautiful here [on the Holy Mountain]—from Holy Pascha until the Panagia's day in August. The beautiful rocks theologize like voiceless theologians, as does all of nature—each creature with its own voice or its silence.[18]

Nor is this love of nature's beauty confined to the Mount Athos. The "face" of God, writes St. Nikolai Velimirovich from Lake Ohrid, "pours beauty over all creation. The universe swims in [God's] beauty as a boat swims in the sea."[19]

"When I began to pray with all my heart," recalls the Russian Pilgrim, "all that surrounded me appeared delightful to me: the trees, the grass, the birds, the earth, the air, and the light."[20] "It is remarkable how the human *nous* sees things differently according to its own light," writes St. Peter of Damaskos, "even when these things . . . in themselves remain what they are."[21] For the ascetic who sees most clearly and most deeply, the divine beauty of nature shines forth most vividly, the visible everywhere revealing the invisible. And already by the third century, this "seeing" had been precisely contextualized as the second step in a progression: *katharsis*, the soul's purification; *theōria*, illumination of the soul, by means of which it is able to "see" the extraordinary depths within what has been there unobtrusively all along; and ultimately, *theōsis*, in which we are united to God, divinized by divine grace. His soul now purified by tears of repentance, the young Zosima in Dostoevsky's *Karamazov* announces his sudden ability to see the beauty of nature: "look at the divine gifts around us: the clear sky, the fresh air, the tender grass, the birds, nature is beautiful and sinless, and we, we alone, are godless and foolish, and do not understand that life is paradise, for we need only wish to understand, and it will come at once in all its beauty."[22]

Theōria physikē, contemplation of nature, enables the purified soul to see the holy beauty of creation. It is central to the ascetic literature of patristic Christianity, which emphasizes that we must first see God in the beauty of creation *before* we can proceed to the ineffable realm of *theōsis*. We must begin with the beauty of creation, just as surely as God's self-revelation first began with nature before proceeding to written scripture. Yet this is just the aesthetic of the Orthodox Church itself: in its icons and liturgy, in the fragrance of the incense and the divine beauty of its hymnology—visible and audible and tangible things that abound and overflow with the unseen, the unheard, the intangible. Everywhere the visible becomes saturated with the invisible, spills over with divine fragrance. Nor is it accidental that the icon, intended to be a visible window into invisible orders, to put us in touch with noetic realities, is at the same time invariably beautiful: an occasion for wonder, since the icon is not primarily meant to be beautiful at all, but simply to present and sustain within the visible what is holy, venerable, invisible. The holy and beautiful arrive together.

Florensky argued that the love of nature, the sense for nature, the susceptibility and vulnerability for—and the appreciation of—its beauty, was bound up with the sense that creation was real, existed on its own, possessed a weighty interiority mirroring a unique relation to the creator. And the shining forth of this depth is precisely its beauty. This manifest interiority, accessible to the purified soul, is what St. Maximos called the inner *logos* of each being, every leaf and stone and animal and person reflecting its own relation to God, its own inherence in the Eternal Logos. And in Byzantine Christianity, this insight is connected historically to a new and radical aesthetic that was to become foundational for the appreciation of nature's beauty and upon which later generations have drawn deeply to the present day, usually without suspecting its origins in ancient Christianity. This new aesthetic was outlined suggestively in a book called *Byzantine Aesthetics* by Dominican Gervase Mathew—Oxford lecturer in Byzantine studies, and "Inklings" member along with Tolkien and Lewis. Mathew argues that Byzantine aesthetics preserved and continued the "surface aesthetics" of antiquity—oriented toward "the intelligibility of nature," carried out through an "essentially mathematical [and ultimately geometrical] approach to beauty [along with] an absorbed interest in optics." But to this it added a "depth aesthetics," which saw visible beauty (the surface aesthetics) as itself the outer expression of inner, invisible realities.[23] Thus, the beautiful—either in nature or in art—could be contemplated on two levels: either as "beauty rendered visible," or more deeply as "the beauty thus reflected."[24]

This depth aesthetic was in part inspired by a broader metaphysical revolution in late antiquity, beginning with Middle Platonism, which saw in visible order and symmetry the manifestation of inward life. But to this, Byzantine Christianity brought a radical affirmation of the reality of the visible, drawn from three principal sources: first, its conviction that the Divine Liturgy itself was the visible, material presentation of spiritual realities, a sacred drama that enacted eternal events unfolding in time—not as a merely semiotic nexus of metaphors or tropes leading away from nature, but as divine gifts rendered within material nature, the tangible communion bread just as patently material as the touch of the Savior's hand. Second, its fundamental conviction that the Incarnation had sacralized the material world, lending it an inherent spiritual dimension. And finally, the Alexandrine school of hermeneutics had taught it to see the visible, tangible, concrete events narrated in scripture as manifestations of noetic realities, yet in a way that did not negate their material reality—i.e., that did not reduce persons and events in scripture to figures and fables, mere allegories, but esteemed them as concrete, material realities, which nevertheless possessed a deeper, inner, mystical dimension. All three factors, then, were richly articulated in the Byzantine East—even as they became diluted in the West, as it gradually loses its taste for sacred art altogether, a process culminating in the merely "religious" art of the Renaissance. Beauty in nature, both that of *physis* and *technē*, is rather the manifest infusion of material nature with the spiritual reality in which it is rooted, revealing how material nature is both a gift (regarded perceptually) and at the same time a manifestation of the giver (regarded spiritually or noetically). And this view, in turn, provides the basis for the *theōria physikē*, the contemplation of noetic realties in created nature, that is articulated in the ascetic tradition from Evagrios to Maximos, where the mystical reading of scripture is linked to the noetic contemplation of *logoi* in nature—rejecting both a fundamentalism of the senses as well as a parallel fundamentalism of the text—and basing both upon the inherence of nature in an Eternal Logos, whose transformative embodiment has now become not just explicit, but the new, true *axis mundi*. I will here only suggest what I have argued elsewhere in this book, that the modern genre of nature writing (from Thoreau, Emerson, and Muir to the present) which has been critical in shaping environmental sensibilities, and which characteristically discerns in the beauty of nature the traces of a hidden holiness that it struggles to articulate, presupposes this spiritual landscape that was first, and far more articulately, developed in ancient Christianity.

III

Are we, then, entitled to conclude that the holy is at the same time the beautiful, and that beauty is hence a sign—perhaps the cardinal sign—that it is, in fact, the holy we have encountered? And may we gather as well that authentic Christianity, traditional and patristic Christianity, preserved and perpetuated as a living tradition in Orthodox Christianity, so far from dismissing the beauty of nature is alone able to adequately perceive it, comprehend it, and properly celebrate it. Whatever relevance the criticisms of Nietzsche and his successors may carry toward other modes of religion, in relation to Orthodox spirituality they would be simply misplaced, if not altogether absurd.

Yet there still remains one final objection to be met. For alongside the celebration of the beauty of nature in the ascetic writings of Byzantine Christianity, one finds at the same time warnings of the *dangers* of this beauty, *caveats* that seem to lend support to Nietzsche's charges that Christianity denigrates the beauty of creation. Cautions concerning the beauty of creation appear throughout the literature of Orthodox asceticism, but perhaps their most powerful formulation can already be found in the Pentateuch. In the fourth chapter of Deuteronomy (15–19), after warning against fabricating idolatrous reproductions of beasts and birds, creeping things on the earth and fish in the water, Moses adds another warning that reveals the true character of idolatry itself: "And *take good heed to your hearts* . . . lest having looked up into the sky, and having seen the sun and the moon and the stars, and all the heavenly bodies, thou shouldest go astray and worship them, and serve them . . . " Don't carve idols, he cautions, but also—and perhaps most importantly—be careful when you look up into the sky and gaze upon a beauty not made by human hands at all, the beauty of the sky that is perhaps the very epitome of all natural beauty.

This suggests that idolatry consists not only, or even primarily, in the making (*technē*) of graven images, and their subsequent worship, but above all in a certain corruption of the heart that leads to an idolatrous gaze—a gaze that is especially vulnerable to being captured by the beauty of creation, here exemplified by the beauty of the heavens that wields such power to naturally elevate the heart toward God, and to which the Psalmist so often refers us. This beauty sets up a longing, a yearning, an *erōs* that moves the soul. It brings the soul into a state of wonder, which for Aristotle was the beginning of all philosophy. But at the same time, it can hold us captive. "Adam used the senses wrongly," argues St. Theodore, a seventh-century Syrian monastic who is drawing upon both Evagrios and Maximos, "and was spellbound [*thaumazō*] by sensory beauty [*to aisthetoi kallos*]." Instead, he explains, "when perceiving the beauty of creatures, he should have referred it to its source and as a consequence have found his enjoyment and his wonder [*thaumazō*] fulfilled in that, thus giving himself a twofold reason for marveling [*thaumazō*] at the Creator," i.e., finding wonder in *both* the sensory and noetic beauties revealed, rather than simply the former.[25] This is the same point made by St. Athanasios, and discussed in Chapter 11, concerning the genesis of idolatry in a refusal to "see through" the sensory to the noetic, alongside a retrograde movement back into the desires, that short-circuits this noetic referral, fixing the awareness solely upon sensory reality.

But beyond this, beauty and the yearning that it engenders bind the soul to that for which it longs. The more powerful the beauty, the more powerful the longing, and the more powerful the bonds that are forged. "Beauty summons all things to itself," writes St. Dionysios, "and gathers everything to itself" (701D). And the divine beauty, which is beyond beings, and thus can render *from itself* to beautiful beings the beauty they possess, generates a divine longing (*erōs*) that leads the soul beyond itself so that it belongs to that for which it longs (701C, 712A). Moreover, as St. Gregory of Nyssa points out in commenting on the *Song of Songs*, the experience of this divine longing renders the soul insatiable: "Even as now the soul that is joined to God is not satiated by her enjoyment of him, so too the more abundantly she is filled up with his beauty, the more vehemently her longings abound."[26] Nor is this provocative beauty to be found only in the starry skies, for all around us we can find "some slight trace of the divine fragrance, which the whole creation, after the manner of a jar for ointments, imitates within itself by the wonders that are seen within it."[27]

Florensky argues that the pagan worship of many gods was sustained through fear. But does it *originate* in fear as well? How is it that paganism in particular, and idolatry as a whole, are first generated? The "Wisdom of Solomon," looking out upon the idolatrous worship of the Hellenistic world, provides a negative answer. Seeing "fire, or wind, or the swift air, or the circle of stars, or the violent water, or the lights of heaven," people have foolishly "delighted [in their] beauty," thereby mistaking them for "the gods which govern the world." This is folly, the Wisdom Book explains, for "by the greatness and beauty of creatures proportionately the maker of them is seen."[28] *Seen*, it must be emphasized, and not inferred, as proponents of the so-called argument from design presuppose. The divine beauty that is invisible somehow, wonderfully, becomes visible in the beauty of creation. So why does the idolater fail to see what is evident? The Wisdom Book answers this question too: It is because of the corruption of the heart, through which "men have lived dissolutely and unrighteously," that they have attached themselves to the beauty of creation rather than recognizing in the beauty of creation the greater, truer, more fulfilling beauty of the Creator.

"Blessed are the pure in heart, for they shall see God." Conversely, the impure in heart—those who do not "take good heed" (*nēpsis*) of their hearts, those whose hearts are corrupted—will to that same extent not see the Creator at all, but rather creatures alone, and thus attach to created things themselves the soul's longing that creation's beauty evokes. In the text already cited, traditionally attributed to St. Theodore the Ascetic, it is argued that the Fall itself resulted from a warp, a distortion or deformation of the way that the *prōtē anthrōpos*, the first man (Adam) looked upon the perceptual world around him, and most especially upon how he responded to its most engaging feature (its beauty) and how he directed his natural, noetic response (his wonder) to that beauty.[29] Somehow, perversely, Adam's wonder was directed not toward the Creator from whom all this beauty proceeds and to whom it testifies, but rather his wonder and longing became fixated upon the sensuous character of the beauty itself—clinging to the message, and ignoring the messenger, and indeed somehow oblivious to the very fact that this wonderful beauty was itself a message at all—a gift and a blessing, and not just a pleasure with which to satisfy ones desires. Rather than his soul (his *nous* or mind, as the highest part of his soul) continuing to ascend from the enticingly beautiful character of the visible to the far higher beauty of the Creator, whose icon he found everywhere around him, the first man somehow became disoriented, confusing instead this higher beauty with what he calls its "bastard offspring"—the all-too-familiar enticements of power and wealth, the self-entitled indulgence in the sensual and visible as ends in themselves. This was the primal idolatry, the inner dynamic of the Fall itself, and to one extent or another we are all, like Adam, idolaters, fixedly bedazzled by, obsessively avaricious of, chronically longing for the purely sensuous beauty of nature, rather than freely wondering at where this could possibly have originated, thereby ascending naturally toward its source. Virtually all modern marketing (it is worth noting) depends upon the sensual wonder and enticement following from the cheap and facile substitution of the immediately visible for its less obvious—and indeed, invisible—but truly life-giving source.

Thus, in authentic, patristic Christianity, we are cautioned about created beauty not in order to deter us from it, nor even less to cultivate contempt for it, but to exhort us to pursue it more deeply, more radically, more authentically to its very source—and thus to warn us against getting side-tracked in the swift currents and swirling eddies of the material world, with its peculiar sorts of back-alleys. And this holds true not just for the beauty of nature, but for the beauty of icons and chant and liturgies and vestments as well.

All these visible beauties are traces of divine fragrance, "droplets of divine love," to which dissolute and unrighteous souls will idolatrously tend to attach themselves, but through which instead the original, and originating, beauty of the Creator can be seen by those whose hearts have been purified.

"We shall be like [God]," writes St. John the Evangelist, "for we shall see Him as He is. And everyone who has this hope in [God] purifies himself, just as [God Himself] is pure" (1 John 3: 2, 3). To see God is to be like God, and thus to be likewise purified. The Evangelist, of course, is writing here specifically about "the time when God appears," about "His coming," "when He is revealed" (1 John 2:28, 3:2). It is truly then that "we shall see Him as he is." To see God in nature truly is to see God, but it is also to see Him (as St. Paul put it) "through a glass darkly," not yet "as He is." Nevertheless, to see God as He has revealed Himself in the beauty of creation is to see Him as he intended Himself to be seen, to see God through one of the principal ways that He chose to manifest Himself. Thus it requires of us that even now, through purifying our souls, we must "be like Him" to the same extent that we would be able to see him, even becoming "partakers" or participants "of the Divine Nature" (2 Peter 1:4) in the very act of seeing, in the scent of "divine fragrance," in the savor of the "droplets of divine love," and in the long-neglected capacity to hear "the voice of the Lord God walking in the Garden" during the afternoon and even to sense His "face" or countenance (*prosōpon*) "in the midst of the trees of the Garden" (Gen 3:8).

Notes

Preface

1. On the city of Constantinople, see Chapter 4 of this book.

2. Ecumenical Patriarch Bartholomew, Foreword, in Margaret Barker, *Creation: A Biblical Vision for the Environment* (London: T & T Clark, 2010), ix.

3. Photographs of this same landscape were never missing from the copies of *Arizona Highways,* stacks of which were obligatory in every physician's waiting room.

4. Bartholomew, "Foreword" to *Creation: A Biblical Vision for the Environment.*

5. Max Oelschlaeger, *Caring for Creation: An Ecumenical Approach to the Environmental Crisis* (New Haven: Yale University Press, 1994).

6. Bruce V. Foltz, *Inhabiting the Earth: Heidegger, Environmental Ethics, and the Metaphysics of Nature* (Atlantic Highlands, N.J.: Humanities Press International, 1995).

7. Michael P. Cohen, *The Pathless Way: John Muir and American Wilderness* (Madison: University of Wisconsin Press, 1984), 281ff. Cohen does a fine job here of showing the radical difference between the genuinely religious consciousness of Muir and its reduction to "a kind of behavioral or subjective experience" that can only exercise itself through "an organizational and institutional form," i.e. through "a sacrifice of spiritual truth for political power."

8. Foltz, *Inhabiting the Earth.*

9. Ibid.

10. "But why must 'the holy' be the poet's word? Because the one who stands 'under favorable weather' has solely to *name* that to which he belongs by virtue of his divining, that is, nature. In awakening, *nature reveals her own essence as the holy.*" And later, Heidegger adds more simply, although it must be emphasized, only by way of exegesis: "*The holy is the essence of nature.*" Martin Heidegger, "As When on a Holiday . . ." in

Martin Heidegger, *Elucidations of Hölderlin's Poetry*, trans. Keith Hoeller (Amherst, N.Y.: Humanity Books, 2000), 81f. Italics added.

11. See, for example, Martin Heidegger, "Building Dwelling Thinking," in Martin Heidegger, *Poetry, Language, Thought*, trans. Albert Hofstadter (New York: Harper & Row, 1971).

12. John Muir, *My First Summer in the Sierra* (Boston: Houghton Mifflin, 1944), 16.

13. *The Pilgrim's Tale*, ed. Aleksei Pentkovsky, tr. T. Allan Smith (New York: Paulist Press, 1999), 77.

14. "What a strange leap, presumably bringing us the insight that we do not yet sufficiently reside where we genuinely are already." Martin Heidegger, "The Principle of Identity," in Martin Heidegger, *Identity and Difference*, trans. Joan Stanbaugh (New York: Harper & Row, 1969), 33; translation modified.

Introduction: The Noetics of Nature

1. Charles Taylor, *A Secular Age* (Cambridge, Mass.: Harvard University Press, 2007), 302.

2. Martin Heidegger, *Sojourns: The Journey to Greece*, trans. John Panteleimon Manoussakis (Albany: SUNY Press, 2005), 26f.

3. Christos Yannaras, *Postmodern Metaphysics*, trans. Norman Russell (Brookline, Mass.: Holy Cross Orthodox Press, 2004), 55.

4. See Martin Heidegger, "Science and Reflection," in Martin Heidegger, *The Question Concerning Technology and Other Essays*, trans. William Lovitt (New York: Harper & Row, 1977), 157 and throughout.

5. Ibid., 164.

6. Ibid., 166ff.

7. Christopher A. Dustin and Joanna A. Ziegler, *Practicing Morality: Art, Philosophy, and Contemplative Seeing* (New York: Palgrave Macmillan, 2007), 9ff. Italics in original.

8. Martin Heidegger, *Parmenides*, trans. André Schuwer and Richard Rojcewicz (Bloomington: Indiana University Press, 1992), 144ff.

9. Christos Yannaras, *Person and Eros*, trans. Norman Russell (Brookline, Mass.: Holy Cross Orthodox Press, 2007), 189f.

10. William James, *The Varieties of Religious Experience* (New York: Barnes and Noble Classics, 2004), 329.

11. Cf. Evan Brann, *The Music of Plato's Republic* (Philadelphia: Paul Dry Books, 2004) and John Sallis, *Being and Logos: The Way of Platonic Dialogue* (Bloomington: University of Indiana Press, 1996).

12. Since *noetic* refers to the proper exercise of the *nous*, the latter is the operative term, and both are notoriously difficult to define either in Ancient Greek or in Byzantine Greek. "Mind" and "intellect" are each hopelessly misleading for English speakers. The glossary definition appended to the English translation of the *Philokalia*, prepared by several of the finest

Byzantine scholars of the time, probably remains unsurpassed, and is worth quoting at length: *Nous* is defined here as "the highest faculty in man, through which—provided it is purified—he knows God or the inner essences or principles of created things by means of direct apprehension or spiritual perception. Unlike the *dianoia* or reason, from which it must be carefully distinguished, the [*nous*] does not function by formulating abstract concepts and then arguing on this basis to a conclusion reached through deductive reasoning, but it understands divine truth by means of immediate experience, intuition or 'simple cognition' (the term used by St. Isaac the Syrian). The [*nous*] dwells in the 'depths of the soul'; it constitutes the innermost aspect of the heart. . . . The [*nous*] is the organ of contemplation, the 'eye of the heart.'" *The Philokalia, Volume Two*, ed. St. Nikodimos of the Holy Mountain and St. Makarios of Corinth, trans. G. E. H. Palmer, Philip Sherrard, and Kallistos Ware (London: Faber & Faber, 1981), 384.

13. Jonathan Hale, *The Old Way of Seeing: How Architecture Lost Its Magic (And How to Get It Back)* (Boston: Houghton Mifflin, 1994), 164.

14. Ibid., 39; Ralph Waldo Emerson, *Emerson: Essays & Lectures*, ed. Joel Porte (New York: The Library of America, Penguin Books, 1983), 7.

15. Hale, *Old Way of Seeing*, 43.

16. Hale, *Old Way of Seeing*, 38.

17. "The expert knowledge of agriculture developed in the universities, like other such knowledges, is typical of the alien order imposed on a conquered land. We can never produce a native economy, much less a native culture, with this knowledge. It can only make us imperialist invaders of our own country." Wendell Berry, *The Unsettling of America: Culture and Agriculture* (New York: Avon, 1978), 168.

18. James Barr, "Of Metaphysics and Polynesian Navigation," in *Seeing God Everywhere: Essays on Nature and the Sacred*, ed. Barry McDonald (Bloomington, Ind.: World Wisdom, 2003), 161–68.

19. Walter Otto, *The Homeric Gods: The Spiritual Significance of Greek Religion [Die Götter Griechenlands: Das Bild des Göttlichen im Spiegel des griechischen Geistes]*, trans. Moses Hadas (New York: Pantheon, 1954), 10.

20. Ibid., 7.

21. Ibid., 17ff.

22. Ibid., 281ff.

23. "This state and manner of being is a state in which we are not closed up but *open*. Wakefulness is openness—the very openness of a huge open door [as Aristotle suggests in his *Generation of Animals*, B 3, 736 b28]. It is not a state of activity, but rather a state of preparedness, of alertness. This state or manner of being is called in Greek *nous* or *noein*." Jacob Klein, "Aristotle, an Introduction," in Jacob Klein, *Lectures and Essays*, ed. Robert B. Williamson and Elliott Zuckerman (Annapolis, Md.: St. John's College Press, 1985), 190.

In the tradition of Byzantine spirituality, it is taught that the *nous* can function in its proper, "open" manner only when it is cleansed of distracting thoughts (*logismoi*) through *katharsis*, and when this openness is cultivated and maintained by watchfulness or mindfulness (*nēpsis*).

24. Peter Brown, *The Rise of Western Christendom: Triumph and Diversity*, second edition (Malden, Mass.: Blackwell Publishing, 2003), 162ff. See also Béatrice Caseau, "Sacred Landscapes," in *Late Antiquity: A Guide to the Postclassical World*, ed. G. W. Bowerstocck, Peter Brown, and Oleg Grabar (Cambridge, Mass.: Harvard University Press, 1999), 21ff.

25. Pavel Florensky, *The Pillar and Ground of Truth: An Essay in Orthodox Theodicy in Twelve Letters*, trans. Boris Jakim (Princeton: Princeton University Press, 1997), 216, 226–30; 531n514.

26. Those accustomed to interpreting the Greek *energeia* by means of the Scholastic notions of *actus* and *actualitas* will need to take care not to read this much later understanding back into the Byzantine experience in which the concept is rooted if they are to understand it clearly and without anachronism.

27. See, for example, Eric D. Perl, *Theophany: The Neoplatonic Philosophy of Dionysios the Areopagite* (Albany: SUNY Press, 2007). It should be noted here the *symbolon* understood patristically is what brings together (*syn*) into a whole (*holos*) heaven and earth, just as the diaboloical is what divides (*dia*) the visible and invisible into opposing realms.

28. Charles Taylor offers an alternative assessment of the art of the Western Renaissance, arguing of its "realism" that "instead of being read as a turning away from transcendence, should be grasped in a devotional context, as a powerful affirmation of the Incarnation." Charles Taylor, *A Secular Age*, 144. But Taylor's own interpretation of this dramatic and comprehensive shift in aesthetic orientation, away from the visible as transparent to the invisible and toward what he acknowledges as the visible fixtures of "everyday life," would seem tenable only under the assumption of a rather Hegelian understanding of the Incarnation, as the irreversible self-emptying of transcendence into immanence. And indeed, Renaissance art needs to be seen as a degradation of "sacred art" into "religious art," i.e., secular art, art that employs the same techniques common to any other subject matter, that merely happens to have a religious orientation. It thus represents an important step toward the secularization process that finds its philosophical bard in Hegel.

29. Giles Deleuze, *Pure Immanence: Essays on A Life* (New York: Zone Books, 2001), 27ff.

30. Introduction to *After God: Richard Kearney and the Religious Turn in Continental Philosophy*, ed. John Panteleimon Manoussakis (New York: Fordham University Press, 2006), xvii.

31. John Panteleimon Manoussakis, *God After Metaphysics: A Theological Aesthetic* (Bloomington: Indiana University Press, 2007), 34. Manoussakis continues that the "icon of the invisible God" is of course Christ, whereas the present work will pursue the extent to which creation itself is such an icon. But since in the Byzantine tradition, which will

centrally inform the conclusions here, it is precisely Christ, the Eternal Logos, who is made manifest in creation, both the problematics as well as the resolutions are ultimately the same.

32. James K.A. Smith, *Introducing Radical Orthodoxy: Mapping a Post-Secular Theology* (Grand Rapids, Mich.: Baker Academic, 2004), 88.

33. Charles Taylor, *A Secular Age*, 309; Guido Vanheeswijck, "Every Man Has a God or an Idol: René Girard's View of Christianity and Religion," in Peter Jonkers and Ruud Welten, *God in France: Eight Contemporary French Thinkers on God* (Leuven: Peeters, 2005, 94).

34. John Milbank, Catherine Pickstock, and Graham Ward, eds., *Radical Orthodoxy: A New Theology* (London: Routledge, 1999), 4.

35. Manoussakis, *After Metaphysics*, 15, 91.

36. Heidegger, *Parmenides*, 149.

37. "Someone who has experienced theology in its own roots, both the theology of the Christian faith and that of philosophy, would today rather remain silent about God when he is speaking in the realm of thinking." Martin Heidegger, "The Onto-Theo-Logical Constitution of Metaphysics," in Martin Heidegger, *Identity and Difference* (New York: Harper & Row, 1969), 54f.

38. Wendell Berry, *Life Is a Miracle: An Essay against Modern Superstition* (New York: Counterpoint, 2001), 8.

39. Seyyed Hossein Nasr, *Religion and the Order of Nature* (New York: Oxford University Press, 1996), 217.

1. Whence the Depth of Deep Ecology? Natural Beauty and the Eclipse of the Holy

1. Albert Camus, "Helen'e Exile," in *The Myth of Sisyphus and Other Essays*, trans. Justin O'Brien (New York: Vintage Books, 1955), 136.

2. Arne Naess, "Deepness of Questions and the Deep Ecology Movement," in *Deep Ecology for the 21st Century: Readings on the Philosophy and Practice of the New Environmentalism*, ed. George Sessions (Boston: Shambala, 1995), 208.

3. Ibid., 207.

4. Ibid., 206.

5. Ibid., 205.

6. Ibid., 68.

7. John Milbank, "Beauty and the Soul," in *Theological Perspectives on God and Beauty*, ed. John Milbank, Graham Ward, and Edith Wyschograd (Harrisburg, Pa.: Trinity Press, 2003), 2; italics added.

8. Mary Mothersill, "Beauty," in *A Companion to Aesthetics*, ed. David Cooper (Oxford: Blackwell Publishers Ltd., 1996), 46

9. Conversely, the great creation narratives, the testimony of which philosophy has always felt obliged to ignore, all precisely *begin* with wonder at

the self-evident reference of the beauty of the cosmos to some otherness from which it originates.

10. Samuel Taylor Coleridge, *Anima Poetae* in *The Norton Book of Nature Writing*, ed. Robert Finch and John Elder (New York: Norton, 1990), 97.

11. Leo Tolstoy, "What Is Art?" in *The Portable Tolstoy*, ed. John Bayley (New York: Viking, 1978), 827ff. Muir, *My First Summer in the Sierra* (Boston: Houghton Mifflin, 1916), 14.

12. Friedrich Nietzsche, *Ecce Homo*, in *Basic Writings of Nietzsche*, trans. and ed. Walter Kaufmann (New York: The Modern Library, 1968), 751.

13. Ibid., 756.

14. "I do not believe in a future life," said Raskolnikov.

> Svidrigailov sat thinking.
>
> "And what if there are only spiders there, or something of the sort," he said suddenly.
>
> "He's a madman," thought Raskolnikov.
>
> "We keep imagining eternity as an idea that cannot be grasped, something vast, vast! But why must it be vast? Instead of all that, imagine suddenly that there will be one little room there, something like a village bathhouse, covered with soot, with spiders in all the corners, and that's the whole of eternity. I sometimes fancy something of the sort." . . .
>
> A sort of chill came over Raskolnikov at this hideous answer.
>
> Fyodor Dostoevsky, *Crime and Punishment*,
> trans. Richard Pevear and Larissa Volikhonsky
> (New York: Vintage Books, 1993), 289f.

15. Vladimir Solovyov, "The Collapse of the Mediaeval World-Conception," in *A Solovyov Anthology*, ed. S. L. Frank (London: St. Austin Press, 2001), 71.

16. Vladimir Solovyov, "The Idea of Humanity," in *A Solovyov Anthology*, 58.

17. Vladimir Solovyov, "Beauty in Nature," in *A Solovyov Anthology*, 127.

18. Vladimir Solovyov, "The Meaning of Art," in *A Solovyov Anthology*, 145.

19. Martin Heidegger, "As When on a Holiday . . ." in *Elucidations of Hölderlin's Poetry*, trans. Keith Hoeller (Amherst, N.Y.: Humanity Books, 2000), 224, 214.

20. Ibid., 203.

21. Martin Heidegger, "Letter on Humanism," in *Basic Writings*, ed. David Farrell Krell (New York: Harper and Row, 1977), 230.

22. Pavel Florensky, *The Pillar and Ground of the Truth*, trans. Boris Jakim (Princeton: Princeton University Press, 1997), 92.

23. Ibid., 137.

24. Ibid., 253ff.

25. Ibid., 132, 176.

26. Ibid., 200; St Paul Florensky, *Salt of the Earth: A Narrative on the Life of the Elder of Gethsemane Skete, Hieromonk Abba Isidore*, trans. Richard Betts (Platina, Calif.: St. Herman of Alaska Brotherhood, 1999).

27. Joyce Carol Oates, "Against Nature," in *The Nature Reader*, ed. Daniel Halpern and Dan Frank (Hopewell, N.J.: The Ecco Press, 1996), 226.

2. Nature's Other Side: The Demise of Nature and the Phenomenology of Givenness

1. Marcel Gauchet, *The Disenchantment of the World: A Political History of Religion*, trans. Oscar Burge (Princeton: Princeton University Press, 1997), 95.

2. Jean Baudrillard, *The Mirror of Production* (St. Louis: Telos, 1981).

3. Jean-Luc Marion, *God Without Being*, trans. Thomas A. Carlson (Chicago: The University of Chicago Press, 1991), 17.

4. Martin Heidegger, "The Onto-Theo-Logical Constitution of Metaphysics," in *Identity and Difference*, trans. Joan Stambaugh (New York: Harper & Row, 1969), 72.

5. Mary Midgley observed, a quarter-century ago: "Man is not adapted to live in a mirror-lined box. . . . We need the vast world, and it must be a world that does not need us; a world constantly capable of surprising us, a world we did not program, since only such a world is the proper object of wonder." *Beast and Man: The Roots of Human Nature* (Ithaca: Cornell University Press, 1978), 362.

6. Erazim Kohak, *The Embers and the Stars: A Philosophical Inquiry into the Moral Sense of Nature* (Chicago: The University of Chicago Press, 1984).

7. Hans-Georg Gadamer, "Heidegger's Later Philosophy," in *Philosophical Hermeneutics*, trans. David E. Linge (Berkeley: University of California Press, 1976), 227.

8. Christopher Banford, ed., *The Noble Traveler: The Life and Writings of O.V. de L. Milosz* (West Stockbridge, Mass.: The Lindisfarne Press, 1985), 332.

9. David Abram, *The Spell of the Sensuous: Perception and Language in a More-Than-Human World* (New York: Pantheon, 1996).

10. See Martin Buber, *I and Thou*, trans. Walter Kaufmann (New York: Scribner, 1970), 146f; and Kohak, *Embers*, 128.

11. Max Scheler, *The Nature of Sympathy*, trans. Peter Heath (Hamden, Conn.: Archon Books, 1973), 82.

12. Pavel Florensky, *The Pillar and Ground of Truth: An Essay in Orthodox Theodicy in Twelve Letters*, trans. Boris Jakim (Princeton: Princeton University Press, 1997), 253.

13. And as noted in the Introduction to the present book, this one-sidedly materialist view was pushed to its very limits, if not modified althogether,

by Lucretius through his restoration of at least the interiority of beauty to the nature of things.

14. "The language . . . is no longer (or not yet) a speaking *about*. It speaks *to* an unknown and nameless God. . . ." Ilse N. Bulhof and Laurens ten Kate. "Echoes of an Embarrassment: Philosophical Perspectives on Negative Theology—an Introduction," in *Flight of the Gods: Philosophical Perspectives on Negative Theology*, ed. Ilse N. Bulhof and Laurens ten Kate (New York: Fordham University Press, 2000), 56f.

15. Czeslaw Milosz, "Advice," in *Poetry for the Earth*, ed. Sara Dunn and Alan Scholefield (New York: Ballantine Books, 1991), 139. Czeslaw Milosz, the Nobel Laureate, describes himself as a "distant cousin" of the elder Oskar Milosz, cited previously, whom he met several times, and whom he describes as "a newcomer" to Paris, arriving from the Polish-Lithuanian border country: "for the French imagination a mythical land inhabited by bears and bearlike human creatures or part-time beasts." Czeslaw Milosz, "Introduction," in Banford, 15.

16. Max Scheler, *On the Eternal in Man*, trans. Bernard Noble (London: SCM Press, 1960), 272f.

17. Annie Dillard, *Pilgrim at Tinker Creek* (New York: Harper Collins, 1974).

18. Heidegger examined at length the "reciprocal conditioning" or "necessary interplay between subjectivism and objectivism." See, *inter alia*, "The Age of the World Picture," in the recent translation of *Holzwege*: Martin Heidegger, *Off the Beaten Track*, trans. Julian Young and Kenneth Haynes (Cambridge: Cambridge University Press, 2002), 66.

19. Wisdom XII, 28–XIII, 10, and Romans I, 29–25.

20. Martin Heidegger, "As When on a Holiday . . ." in *Elucidations of Hölderlin's Poetry*, trans. Keith Hoeller (Amherst, N.Y.: Humanity Books, 2000). For a discussion of these modes of approaching nature in Heidegger's thought, see Bruce V. Foltz, *Inhabiting the Earth: Heidegger, Environmental Ethics, and the Metaphysics of Nature* (Englewood, N.J.: Humanities Press, 1995).

21. The concept of iconic vision is discussed at length in Chapter 6, "The Iconic Earth: Nature Godly and Beautiful."

22. Marion, *God Without Being*, 76.

23. Martin Heidegger, *Vorträge und Aufsätze* (Pfullingen: Neske, 1978), 31f.

24. Friedrich Nietzsche, *The Gay Science*, trans. Walter Kaufmann (New York: Random House, 1974), 181.

25. Carolyn Merchant, *The Death of Nature: Women, Ecology, and the Scientific Revolution* (New York: Harper & Row, 1980), 168–72. Donne as cited in Albert Borgmann, *Crossing the Postmodern Divide* (Chicago: The University of Chicago Press, 1992), 22, emphasis added.

26. For a powerful argument that for modern medicine, the human body already *is* a corpse, see Jeffry P. Bishop, *The Anticipatory Corpse: Medicine,*

Power, and the Care of the Dying (South Bend, Ind.: The University of Notre Dame Press, 2011).

3. Layers of Nature in Thomas Traherne and John Muir: Numinous Beauty, Onto-theology, and the Polyphony of Tradition

1. Martin Heidegger, *Die Kategorien- und Bedeutungslehre des Duns Scotus*, in *Frühe Schriften* (Frankfurt am Main: Klostermann, 1972), 352.
2. Ibid., 147, 344.
3. On the cancellation of the mysticism course, see Theodore Kisiel, *The Genesis of Heidegger's "Being and Time"* (Berkeley: University of California Press, 1993), 76. On the circumstances of Heidegger's introduction to the mysticism of the Latin Middle Ages, John van Buren notes: "Joseph Sauer's course on 'The History of Medieval Mysticism,' which Heidegger had taken in WS 1910–1911, probably sparked his lifelong preoccupation with mysticism and especially Meister Eckhart." John van Buren, *The Young Heidegger: Rumor of the Hidden King* (Bloomington: University of Indiana Press, 1994), 62f.
4. John D. Caputo, *The Mystical Element in Heidegger's Thought* (Athens: Ohio University Press), 153.
5. For example, in the WS 1920–21 course on St. Paul, "Introduction to the Phenomenology of Religion," Heidegger already suggests an equation of mysticism with "Hellenistic mystery-religions," and goes on to state a dramatic contrast between what he calls "the Mystics" and "the Christians": "The Mystic is, through manipulation, removed from the life-complex; in an enraptured state God and the universe are possessed. The Christian knows no such 'enthusiasm,' rather he says: 'let us be awake and sober.' Here precisely is shown to him the terrible difficulty of the Christian life." Martin Heidegger, *The Phenomenology of Religious Life* (Bloomington: University of Indiana Press, 2004), 88f.
6. By the end of 1917, Heidegger had married his Lutheran student Elfriede Petri. He had spent the following summer reading Schleiermacher, and by the end of the year Husserl had noted Heidegger's "radical" break with Catholic doctrine and his "migration" to Protestantism.
7. Martin Heidegger, "The Problem of Sin in Luther," in *Supplements: From the Earliest Essays to "Being and Time" and Beyond*, trans. John van Buren (Albany: SUNY Press, 2002), 107. This is taken from the talk given by Heidegger in Bultmann's seminar on St. Paul at Marburg, in 1924.
8. Heidegger, *Phenomenology of Religious Life*, 213.
9. Martin Luther, *Theologie des Kreuzes*, cited in van Buren, *The Young Heidegger*, 188.
10. Ibid. Italics in original.
11. Heidegger, *Phenomenology of Religious Life*, 70.
12. Ibid., 73f.
13. Ibid., 110f, 67.

14. Giani Vattimo, "Heidegger and Christian Existence," in Vattimo, *After Christianity*, trans. Luca D'Isanto (New York: Columbia University Press, 2002), 123f. Italics in original.

15. Martin Heidegger, *The Metaphysical Foundations of Logic*, trans. Michael Heim (Bloomington: Indiana University Press, 1984), 165.

16. "*Der Spiegel* Interview with Martin Heidegger," in *The Heidegger Reader*, ed. Gunter Figal, trans. Jerome Veith (Bloomington: Indiana University Press, 2009), 326. It is clear from Heidegger's later writings, especially some of the unpublished texts, that he came to believe a viable, living encounter with the divine is not an option for the foreseeable future. See, for example, *Besinnung* (translated as *Mindfulness*), written soon after his *Beitrage*, where he flatly states that in our current epoch, "gods have become impossible." Martin Heidegger, *Mindfulness* (London: Continuum, 2006), 208ff.

17. Thomas Traherne, *Christian Ethicks*, Chapter XXXII, "Of Gratitude," in *Thomas Traherne: Poetry and Prose*, ed. Denise Inge (London: SPCK, 2002), 69.

18. Traherne, *Centuries of Mediations*, "The First Century, in Inge, 3ff.

19. Traherne, *Select Meditations*, II.71, in Inge, 85.

20. Traherne, *Centuries*, "Second Century," in Inge, 12.

21. Traherne, *Christian Ethicks*, Chapter XXVI, "Humility," in *The Way to Blessedness: Thomas Traherne's Christian Ethicks*, The Spelling and Punctuation Modernized, ed. Margaret Bottrall (London: The Faith Press, 1962), 231.

22. Ibid., 230.

23. S. Sandbank, "Thomas Traherne on the Place of Man in the Universe," *Scripta Hierosolymitana* 17 (1966): 122, cited in Donald R. Dickson, *The Fountain of Living Waters: The Typology of the Waters of Life in Herbert, Vaughan, and Traherne* (Columbia: University of Missouri Press, 1987), 173.

24. Traherne, cited in Dickson, *Fountain of Living Waters*, 173.

25. Robert Watson characterizes Traherne's "childlike vision of nature" as "open, intense, full, and unthreatened"—as being "so vivid that the ego nearly dissolves into the given universe." Yet rather than seeing this relation to the natural world as radically transcending of the spheres of both material self-interest (Hobbes) and the reflexivity of the *cogito* (Descartes), he misconstrues it as a Berkeleian subjectivism, which leads him to designate Traherne's mysticism "in-static" rather than "ecstatic," even as he admits that this is "a distinction almost without a difference." And apparently unaware of the ancient patristic teaching that the visible universe was created to serve as an image revealing its Creator, he remarks sardonically that Traherne's God is "a bit like Santa Claus in his workshop, creating toys for the joy of the human child and taking pleasure in the way the children enjoy them." As if a proper deity, one with more gravitas, would never condescend to create a universe so delightful to its inhabitants. All this leads him to conclude that "because the very purpose of that universe is to serve

and delight the individual soul, Traherne's religious psychology [*sic*] ultimately rejects any version of the Eastern apophatic theologies by negations of human categories, by blanking out of what the human mind can hold." (But, of course, the very goal of apophatic theology in the Eastern tradition is to trace a path upon which the soul can arrive at a direct experience of God, the first stages of which [*theōria physikē*] are quite similar to what Traherne describes.) Thus, despite an evident appreciation for Traherne's view of nature, irony and cynicism carry the day: "Objects thus have a glorious presence and plenitude for Traherne, in a way that challenges the system I have been describing—at least by complicating it, and perhaps by offering an alternative path that now looks merely like a quaint dead-end country lane only because Western intellectuals have driven so far on the other, materialist-pessimist freeway." It is hard not to hear in the wit of this passage the echo of a certain melancholy. Robert N. Watson, *Back to Nature: the Green and the Real in the Late Renaissance* (Philadelphia: University of Pennsylvania Press, 2007), 297ff; italics added.

26. Traherne, "Dumness," in *Thomas Traherne: Centuries, Poems, and Thanksgivings*, vol. II, ed. H. M. Margoliouth (Oxford: Oxford University Press, 1958), 41f.

27. Although the earlier appraisal, once associated with Roderick Nash, still lingers that Muir was, as Worster puts it, the apostle of "a post-Christian religion of nature," as previously unpublished papers are being assimilated into the Muir canon and newer scholarship emerges, it is becoming clearer that his religious beliefs were quite consonant with much of the Protestant religious thought of his time. Donald Worster, *A Passion for Nature: The Life of John Muir* (Oxford: Oxford University Press, 2008), 306. In contrast, see especially Dennis C. Williams, *God's Wilds: John Muir's Vision of Nature* (College Station: Texas A&M Press, 2002). For a briefer expression of the reappraisal, see John Gatta, in *Making Nature Sacred: Literature, Religion, and Environment in America from the Puritans to the Present* (Oxford: Oxford University Press, 2004), 148–57, which maintains that Muir's "robust piety drew constantly on biblical paradigms of grace, conversion of heart, evangelical poverty, and a loving Creator" (Ibid., 150).

28. John Muir, *A Thousand-Mile Walk to the Gulf*, ed. William Frederic Badè (Boston: Houghton Mifflin, 1941), 30.

29. Ibid., 41f.

30. Ibid., 98.

31. Ibid., 69.

32. John Muir, *The Cruise of the Corwin: Journal of the Arctic Expedition of 1881 in Search of DeLong and the Jeannette*, ed. William Frederic Badè, in *John Muir: His Life and Letters and Other Writings*, ed. Terry Gifford (London: Bâton Wicks, 1996), 746.

33. John Muir, *Our National Parks*, in *The Eight Wilderness Discovery Books* (Seattle: The Mountaineers, 2004), 490.

34. John Muir, *My First Summer in the Sierra* (Boston: Houghton Mifflin, 1944), 213.
35. Chris Highland, Introduction to *Meditations of John Muir: Nature's Temple* (Berkeley, Calif.: Wilderness Press, 2001), ix.
36. Michael P. Cohen, *The Pathless Way: John Muir and American Wilderness* (Madison: University of Wisconsin Press, 1984), 150, 282.
37. Cited in Williams, *God's Wilds*, 7.
38. John McPhee, *Coming into the Country* (New York: Farrar, Straus and Giroux, 1991).
39. Wendell Berry, "An Entrance to the Woods," in *Recollected Essays: 1965–1980* (San Francisco: North Point Press, 1981), 233–36.
40. Traherne, *Centuries*, 3.5, in Inge, 18.
41. Fyodor Dostoevsky, *The Brothers Karamazov*, trans. Richard Pevear and Larissa Volokhonsky (New York: Vintage Books, 1991), 319.
42. Lynn White Jr, "The Historical Roots of Our Ecologic Crisis," in *Dynamo and Virgin Reconsidered: Essays in the Dynamism of Western Culture* (Cambridge: MIT Press, 1971), 88.
43. Ibid., 11.

4. Sailing to Byzantium: Nature and City in the Greek East

An earlier version of this chapter was presented at the "Natural City" conference, held in June 2004 at the University of Toronto.

1. Jane Taylor, *Imperial Istanbul: A Traveller's Guide* (London: I. B. Tauris, 1998), 14.
2. W. B. Yeats, *A Vision* (New York: Macmillan, 1961), 279f.
3. W. B. Yeats, Preface to *The Works of William Blake: Poetic, Symbolic, and Critical.* Ed. with Lithographs of the Illustrated Prophetic Books and a Memoir and Interpretation by E. J. Ellis and W. B. Yeats, 3 vols. (London: Quaritch, 1893); italics added.
4. G. P. Fedotov, *The Russian Religious Mind. Vol. I: Kievan Christianity: the 10th to the 13th Centuries* (New York: Harper & Brothers, 1960), 372.
5. Martin Heidegger, *Elucidations of Hölderlin's Poetry*, trans. Keith Hoeller (Amherst, N.Y.: Humanity Books, 2000), 185.
6. Ibid., 156.
7. Ibid., 76.
8. Vincent Scully, *Architecture: The Natural and the Manmade* (New York: St. Martin's Press, 1999), 6ff.
9. Ibid., 99–121.
10. See Nietzsche's *Birth of Tragedy*, Wörringer's *Abstraction and Empathy*, and Heidegger's *Origin of the Work of Art.*
11. See the article by H. Liebeschütz on "Frankish Criticism of Byzantine Theories of Art," in *The Cambridge History of Later Greek & Early*

Medieval Philosophy, ed. A. H. Armstrong (Cambridge, Mass.: Cambridge University Press, 1970), 565–75.

12. St. Maximos the Confessor, *Ambigua*, PG 91, 1148C, trans. in Paul Evdokimov, *The Art of the Icon: A Theology of Beauty*, Fr. Steven Bigham, trans. (Redondo Beach, Calif.: Oakwood Publications, 1990), 12.

13. Christopher Tadgell, *Imperial Space: Rome, Constantinople and the Early Church* (London: Ellipsis, 1998), 214.

14. W. Eugene Kleinbauer, *Saint Sophia at Constantinople: Singulariter in Mundo*, Frederic Lindley Morgan Chair of Architectural Design Monograph No. 5 (Louisville: Allen R. Hite Art Institute, 1999), 40.

15. Ibid., 40.

16. Béatrice Caseau, "Sacred Landscapes," in *Late Antiquity: A Guide to the Postclassical World*, ed. G. W. Bowersock, Peter Brown, and Oleg Grabar (Cambridge, Mass.: Harvard University Press, 1999), 21ff.

17. Peter Brown, *The Rise of Western Christendom: Triumph and Diversity*, second edition (Malden, Mass.: Blackwell Publishing, 2003), 164.

18. Ibid., 172ff.

19. Alexandros Papadiamandis, *The Boundless Garden: Selected Short Stories*, Volume I (Limni, Evia, Greece: Denise Harvey Publisher, 2007).

20. See Heidegger's essay, "Building Dwelling Thinking," in Martin Heidegger, *Poetry, Language, Thought*, trans. Albert Hofstadter (New York: Harper & Row, 1971), 141 passim. See also Bruce V. Foltz, *Inhabiting the Earth: Heidegger, Environmental Ethics, and the Metaphysics of Nature* (Englewood, N.J.: Humanities Press, 1995), 154 passim.

5. The Resurrection of Nature: Environmental Metaphysics in Sergei Bulgakov's *Philosophy of Economy*

1. Michael Crichton, "Environmentalism as Religion," delivered September 15, 2003 to the Commonwealth Club in San Francisco; widely cited online, for example: http://scienceandpublicpolicy.org/images/stories/papers/commentaries/crichton_3.pdf.

2. George Sessions ed., *Deep Ecology for the 21st Century: Readings on the Philosophy and Practice of the New Environmentalism* (Boston: Shambala, 1995), 77.

3. Ibid., 233.

4. Erazim Kohák, *The Embers and the Stars: A Philosophical Inquiry into the Moral Sense of Nature* (Chicago: University of Chicago Press, 1984), 22ff.

5. Henry David Thoreau, *The Portable Thoreau*, ed. Carl Bode (New York: Penguin Books, 1982), 616f.

6. Kohák, 182f.

7. Steven G. Marks, *How Russia Shaped the Modern World: From Art to Anti-Semitism, Ballet to Bolshevism* (Princeton: Princeton University Press, 2003).

8. "The Russian Primary Chronicle," cited in *Medieval Russia: A Source Book, 900–1700*, ed. Basil Dmytryshyn (New York: Holt: Rinehart and Winston, 1967), 34.
9. Sergei Bulgakov, *Philosophy of Economy: The World as Household*, trans. Catherine Evtuhov (New Haven: Yale University Press, 2000), 41.
10. Ibid., 40.
11. Ibid.
12. Ibid., 35, 43.
13. Ibid., 181.
14. Ibid., 132.
15. Ibid., 38.
16. Ibid., 77.
17. Ibid., 85.
18. Ibid., 110f.
19. Ibid., 124f.
20. Ibid., 126.
21. Ibid., 131.
22. Ibid., 132.
23. Ibid., 304n7)
24. Ibid., 130.
25. Ibid., 310n34.
26. Marjorie Hope Nicolson, *Mountain Gloom and Mountain Glory: The Development of the Aesthetics of the Infinite* (New York: Norton, 1963).
27. V. S. Naipaul, *Vintage Naipaul* (New York: Vintage Books, 2004), 55.
28. Sergei Bulgakov, *The Comforter*, trans. Boris Jakim (Grand Rapids: Eerdmans, 2004), 208. I have substituted here the Greek "*logoi*" for the translator's rendering from the Russian as "logoses."
29. Cited in Bulgakov, *Economy*, 137f.
30. Cited in Bulgakov, *Economy*, 305.
31. Ibid., 138; Sergei Bulgakov, "Hagia Sophia," in Sergius Bulgakov, *A Bulgakov Anthology*, ed. Nicholas Zernov and James Pain (Philadelphia: The Westminster Press, 1976), 14.
32. Selection from Sergei Bulgakov, *The Unfading Light*, cited in Sergeii Bulgakov, *Towards a Russian Political Theology*, ed. Rowan Williams (Edinburgh: T&T Clark, 1999), 140.
33. Bulgakov, *Comforter*, 202. Bulgakov continues here: "The beauty of nature is a self evident fact—for both believers and unbelievers equally . . . Beauty is the exteriorized sophianicity of creation that 'clothes' the latter; it is the reflection of the mystical light of the Divine Sophia. The beauty of nature is objective. This means that it can by no means be identified with human emotional or subjective states. The beauty of nature is a spiritual force that testifies about itself to the human spirit."
34. Henry David Thoreau, "Walking," in *The Portable Thoreau*, ed. Carl Bode (New York: Penguin Books, 1982), 621, 625.

35. Bulgakov, *Economy*, 151.
36. Bulgakov, *Comforter*, 202.
37. Bulgakov, *Economy*, 146.
38. Ibid., 154.
39. Sergius Bulgakov, *The Bride of the Lamb*, trans. Boris Jakim (Grand Rapids: Eerdmans, 2002), 178.
40. Bulgakov, *Economy*, 144f.
41. Bulgakov, *Bride*, 178f. The text continues: "The remembrance of an edenic state and of God's garden is nevertheless preserved in the secret recesses of our self-consciousness, as an obscure *anamnēsis* of another [mode of] being, similar to the dreams of golden childhood and most accessible to childhood. These are distinct, palpable revelations of the world's sophianicity in our soul, although they are usually obscured in the soul by our failure to believe in their genuineness or even in their possibility."
42. Bulgakov, *Economy*, 183f.
43. Ibid.
44. Ibid., 157.
45. Ibid., 155.
46. Ibid.
47. Ibid., 135, 156. This passage alludes to Bulgakov's theologically important distinction between the divine Sophia, the Wisdom of God as it is rooted in the Trinitarian Life, and the worldly or cosmic Sophia, that same Wisdom as it is manifest in the world. Many theological criticisms of Bulgakov's "sophiology" have ignored this distinction.
48. Bulgakov, *The Unfading Light*, cited in Sergei Bulgakov, *Sophia: The Wisdom of God* (Hudson, N.Y.: Lindisfarne Press, 1993), ix, and *Bulgakov Anthology*, 10.

6. The Iconic Earth: Nature Godly and Beautiful

The term "*godly*" in the title and elsewhere is meant in the older sense of the word, meaning "what comes from God" (ME: *godlîc*), not in the more recent sense of "pious." It is meant, rather, precisely in the sense of John Muir's happy neologism with which he refers in 1913 to the "*Godful* beauty" of nature, or when he writes in 1902 of how the Grand Canyon's "wildness so *godful*, cosmic, primeval, bestows a new sense of earth's beauty," (John Muir, "The Grand Canyon of the Colorado," *Century Illustrated Magazine* 65 [November, 1902], 107–16), or four years later writing of the same place, exclaims: "It is wild, *godful*, cosmic, primeval place bestowing a new sense" (John Muir, "The Grand Canyon of the Colorado," in *Steep Trails*, ed. William Frederic Bade (Boston: Houghton Mifflin, 1918).

1. Fyodor Dostoevsky, *The Brothers Karamazov*, trans. Richard Pevear and Larissa Volokhonsky (New York: Vintage Books, 1991), 362.
2. St. Isaac of Syria, *Daily Readings with St. Isaac of Syria*, ed. A. M. Allchin, trans. Sebastian Brock (Springfield, Ill.: Templegate, 1989), 29.

3. The usage of the words "theology" and "theological" here implies neither the sense of Aristotle's "first philosophy," nor a particular position in regard to the academic dichotomy of "revealed" and "philosophical" theology, nor least of all (as should be evident) what Heidegger and others have called "onto-theology." For the Eastern Church, theology is an empirical science based upon noetic insights, and it is to this sense the the term here aspires, no doubt in a very rudimentary manner.

4. Along with scattered articles in environmental aesthetics are monographs by Arnold Berleant and a survey by the Finnish philosopher Sepänmaa, while Max Oelschlaeger has written an apologia for religious narratives concerning the environment.

5. Max Oelschlaeger, *Caring for Creation: An Ecumenical Approach to the Environmental Crisis* (New Haven: Yale University Press, 1994), 238.

6. James I. McClintock, "'Pray Without Ceasing': Annie Dillard among the Nature Writers," in *Earthly Words: Essays on Contemporary American Nature and Environmental Writers*, ed. John Cooley (Ann Arbor: The University of Michigan Press, 1994), 69f. The two citations in the quote, to the first of which italics have been added, are from John Hildebidle and from Edward Abbey's *Beyond the Wall.*

7. "Our reason has driven all away. Alone at last, we end up by ruling over a desert. . . . We turn our backs on nature; we are ashamed of beauty." Albert Camus, *The Myth of Sisyphus and Other Essays* (New York: Vintage, 1983), 189.

8. Friedrich Hölderlin, *Essays and Letters on Theory*, trans. Thomas Pfau (Albany: SUNY Press, 1988), 94.

9. Pavel Florensky, *The Pillar and Ground of the Truth*, trans. Boris Jakim (Princeton: Princeton University Press, 1997), 234.

10. Elder Porphyrios, *Wounded by Love: The Life and Wisdom of Elder Porphryius* (Limni, Evia, Greece: Denise Harvey, 2005), 107.

11. Hans-Georg Gadamer, "The Relevance of the Beautiful," in *The Relevance of the Beautiful and Other Essays*, trans. Nicholas Walker (Cambridge: Cambridge University Press, 1986), 31.

12. Kierkegaard's "stages" might then be taken as: aesthetic—ETHICAL—religious.

13. Martin Heidegger, *Identity and Difference*, trans. Joan Stambaugh (NewYork, Harper and Row, 1969), 72.

14. Martin Heidegger, *Hölderlins Hymnen "Germanien" und "Der Rhein."* Gestamtausgabe, II. Abteilung, Band 39, ed. Suzanne Ziegler (Frankfurt: Klostermann, 1980), 195; Martin Heidegger, *Contributions to Philosophy (from Enowning)*, trans. Parvis Emad and Kenneth Maly (Bloomington: Indiana University Press, 1999), 63, 75ff.

15. Heidegger, *Identity and Difference*, 54. (The English translation obscures this point, making Heidegger's exclusion of *theologia* from his critique of onto-theology appear to be an inclusion.)

16. Michel Haar, "Heidegger and the God of Hölderlin," *Research in Phenomenology* XIX (1989), 99. See p. 92: "The God of Hölderlin resembles [the God of] the Old Testament, the God of the Psalms, who governs the forces of nature and at the same time watches over the destiny of men, God who is the bestower of the sun and rain as much as of joy and sadness. By ontologizing them, Heidegger makes an attempt to displace the biblical notions of a personal God, of creation and above all the heavenly gift . . . For him these notions belong to medieval ontotheology. Yet they are clearly older than that, almost breaking free of history because they traverse different epochs [by articulating] a phenomenological content belonging to all religions . . ."

17. Jacques Derrida, "Faith and Knowledge: the Two Sources of 'Religion' at the Limits of Reason Alone," in Jacques Derrida and Gianni Vattimo, *Religion* (Stanford: Stanford University Press, 1998), 12.

18. Lynn White, Jr., "The Historical Roots of Our Ecological Crisis," in *Ecology and Religion in History*, ed. David and Eileen Spring (New York: 1974), 22f.

19. Lynn White, Jr., *Medieval Technology and Social Change* (Oxford: Oxford University Press, 1964), 134.

20. Ibid., 30f.

21. Ibid., 26. Italics added.

22. Haar, 94.

23. Georges Duby, *Medieval Art Vol. I: The Making of the Christian West* (Geneva: Skira, 1995), 179f.

24. Ibid., 180. Italics added.

25. Georges Duby, *Medieval Art Vol. III: Foundations of a New Humanism* (Geneva: Skira, 1995), 12.

26. "Nature never made anything that [Giotto] did not imitate or even reproduce in paint, so that men who see his work are often deluded into taking the painted for the real." Cited in Duby, Vol. III, 200.

27. Ibid., 210, 46.

28. Michel Quenot, *The Icon: Window on the Kingdom*, trans. A Carthusian Monk (Crestwood, N.Y.: St. Vladimir's Seminary Press, 1996), 72.

29. Paul Evdokimov, *The Art of the Icon: A Theology of Beauty*, trans. Fr. Steven Bigham (Redondo Beach, Calif.: Oakwood Publications, 1990), 168.

30. Ibid., 201.

31. Umberto Eco, *Art and Beauty in the Middle Ages* (New Haven: Yale University Press, 1986), 56f.

32. Ibid., 84.

33. Ibid., 89.

34. An even sharper contrast is offered by the 1966 film masterpiece by A. Tartovsky, *Andrei Rublev*, which dramatizes the intense and prolonged spiritual preparation of yet another contemporary of the Florentine masters,

the materially impoverished, sainted Russian monk and iconographer Rublev, who is generally regarded as the greatest iconographer of all.

35. Cited in Evdokimov, 14.

36. St. John of Kronstadt, *My Life in Christ or Moments of Spiritual Serenity and Contemplation, of Reverent Feeling, Of Earnest Self-Amendment and Of Peace in God: Extracts from the Diary of St. John of Kronstadt* (Jordanville, N.Y..: Holy Trinity Monastery, 1994), 421; italics added.

37. "Nature on Athos has a special charm, which is the radiance of prayer and holiness. The uncreated grace indeed passes through the soul to the body and spreads even to nonrational nature, to all creation. Nothing is fierce; everything is calm. All night and all day Athos is consumed in prayer . . . It might be that one sees it through the eyes of the God-bearing monks and becomes illumined. It might be that one sees it not through the eye of the mind but through the deified heart. And the heart knows how to love and how to appreciate things." Metropolitan of Nafpaktos Hierotheos, *A Night in the Desert of the Holy Mountain* (Levadia, Greece: Birth of the Theotokos Monastery, 1991), 30.

38. From the viewpoint of yet another Eastern faith tradition, Abraham Heschel writes: "If God were a theory, the study of theology would be the way to understand Him. But God is alive and in need of love and worship." *God in Search of Man: A Philosophy of Judaism* (New York: Farrar, Straus and Giroux, 1955).

39. St. John of Damascus, *On the Divine Images*, trans. David Anderson (Crestwood, N.Y.: St. Vladimir's Seminary Press, 1997), 43.

40. Ibid., 24.

41. Ibid., 23.

42. The parallels with Heidegger here are manifold. For example, the essence of modern technology—like the fallenness of creation resulting from the fall of humanity—prevents the earth from being earth, i.e. from showing its "divinely beautiful" character in any but an occasional manner. And the *Ereignis* that Heidegger sees as already at work presents a certain parallel with the Orthodox thought of a paradisiacal transfiguration that is now at work in nature, but that remains incomplete. Most important, both deal with the ontology of nature itself—not merely with our experience of it.

43. "If you insult the royal robe, do you not insult him who wears it? Do you not know that if you insult the image of the king, you transfer the insult to the original?" St. John Chrysostom, cited in St. John of Damascus, 68. Translation modified.

44. *The Lenten Triodion*, trans. Mother Mary and Archmandrite Kallistos Ware (South Canaan, Pa.: St. Tikhon's Seminary Press, 1999), 306. For a discussion of this text, and the Seventh Ecumenical Council whose teaching it expresses, see Leonid Ousepensky and Vladimir Lossky, *The Meaning of Icons* (Crestwood, N.Y.: St. Vladimir's Seminary Press, 1982), 31f.

45. Lars Thunberg, *Microcosm and Mediator: The Theological Anthropology of Maximos the Confessor* (Chicago: Open Court, 1995), 74.

46. St. John of Kronstadt, 434f.

47. St. Gregory Palamas, "Topics of Natural and Theological Science," *The Philokalia* IV (London: Faber and Faber, 1995), 417.

48. Vladimir Lossky, *The Mystical Theology of the Eastern Church* (Crestwood, N.Y.: St. Vladimir's Seminary Press, 1976), 101.

49. In his presentation at the 1994 Environmental Seminar, sponsored by the Ecumenical Patriarch Bartholomew at Halki Seminary in Turkey, the Romanian theologian Dimitru Popescu maintains that the modern, deistic "isolation of God in His transcendence," through the notion of autonomous nature, becomes at the same time the foundation for environmental catastrophe because it allows human contrivance to fill in the vacuum: "Eastern Theology has its own authoritative voice to add. It is the only theology capable of overcoming the belief in the autonomy of creation typical of the secularized culture, by asserting paradoxically both God's transcendence opposite creation and His immanence in creation. As the sun is not confused with the earth even though it is permanently present in the life of the earth through its light and heat which make life possible, so God, while He remains in His being beyond the world in unapproachable transcendence, is nevertheless present at the same time in the cosmos through the rays of uncreated energies (rays of light, life and love) by which the world was created and recreated in Christ and destined to become a new heaven and a new earth in Christ in the ages to come. The world is not autonomous, but theonomous, because it comes from God and it returns to God who preserves an inner connection with his creation through His uncreated energies. Otherwise the world would slide back into the nothingness out of which it was brought to light by its Creator." "Toward an Orthodox Ecological Education: Theological and Spiritual Principles," accessible at www.patriarchate.org/visit/html/94_35.html.

50. Cited and translated in Olivier Clément, *The Roots of Christian Mysticism* (London: 1993, New City Press), 219; italics added.

51. Barry Lopez, *About This Life: Journeys on the Threshold of Memory* (New York: Knopf, 1998), 134.

52. For this reason, pilgrimages to monasteries are seen in traditional Orthodoxy not as outings, but more as indispensable workshops with men and women who devote there lives entirely to the prayer and contemplation that is expected of all believers.

53. *De Imaginibus* I.8, cited and translated in Eric D. Perl, ". . . That Man Might Become God": Central Themes in Byzantince Theology," in *Heaven on Earth: Art and the Church in Byzantium*, ed. Linda Safran (University Park: The Pennsylvania State University Press, 1998), 56.

54. Mikel Dufrenne, *The Phenomenology of Aesthetic Experience*, trans. Edward S. Casey, et al (Evanston: Northwestern University Press, 1973), 86–89.

55. Ibid., 88. (Dufrenne is paraphrasing Levinas here.)

56. Ibid., 86, 88.
57. Evdokimov, 221.
58. Bruce V. Foltz, *Inhabiting the Earth: Heidegger, Environmental Ethics, and the Metaphysics of Nature* (Atlantic Highlands, N.J.: Humanities Press, 1995).
59. Eugene Hargrove, "An Ontological Argument for Environmental Ethics," in *Foundations of Environmental Ethics* (Englewood Cliffs, N.J.: Prentice-Hall, 1987), 193; italics added.
60. Henry David Thoreau, *The Maine Woods,* in *The Library of America: Henry David Thoreau,* ed. Robert F. Sayre (New York: Literary Classics of the United States, 1985), 644f.
61. Ibid., 646. Italics in original.
62. Evdokimov (p. 180) notes: "The attitude of the Christian West toward images, on the other hand, as shown in the Council of Trent, puts the accent on anamnēsis, memory, but not on the epiphanic presence."
63. Ibid., 221.
64. Leonid Ouspensky, *Theology of the Icon,* Vol. I, trans. Anthony Gythiel (Crestwood, N.Y.: St. Vladimir's Seminary Press, 1992), 178.
65. Evdokimov, 221.
66. Ibid., 169.
67. Archimandrite Vasileios of Stavrokikita, *Hymn of Entry: Liturgy and Life in the Orthodox Church,* Contemporary Greek Theologians, Vol. I (Crestwood, N.Y.: St. Vladimir's Seminary Press, 1984), 89.
68. Annie Dillard, *Pilgrim at Tinker Creek* (New York: Harper Collins, 1974), 9f.
69. Prayer and worship exhibit the transactional (and radically transitive) structure of such acts as praising, offering-up (sacrificing), awaiting, and thanking.
70. Anna Kartsonis, "The Responding Icon," in *Heaven on Earth: Art and the Church in Byzantium,* 65.
71. Konrad Onasch and Annemarie Schnieper, *Icons: the Fascination and the Reality,* trans. Daniel G. Conklin (New York, Riverside Book Company, 1995), 277.
72. Fr. Steven Bigham, *Heroes of the Icon* (Torrance, Calif.: Oakwood Publications, 1998), 52.
73. St. John of Kronstadt, 422.
74. Vasileios, 82. The citation is from the Holy Friday Vespers, prefigurring the Resurrection.
75. John Muir, *The Mountains of California* (New York: Penguin Books, 1989), 179.
76. Wendell Berry, *Recollected Essays 1965–1980* (San Francisco: North Point Press, 1981), 237.
77. John Baggley, *Doors of Perception: Icons and Their Spiritual Significance* (Crestwood, N.Y.: St. Vladimir's Seminary Press, 1995), 106.
78. Dillard, 12.

79. Dufrenne, 548.

80. "It could even be that *nature, in the face she turns* towards [our] technological usurpation, is simply concealing her essence." Martin Heidegger, "Letter on Humanism," in *Basic Writings*, ed. David Farrell Krell (New York: Harper and Row, 1977), 205; italics added and translation modified. For a discussion of "moods" of nature, see Martin Heidegger, *Hölderlins Hymnen*, 104, cited in Foltz, *Inhabiting the Earth*, 144.

81. Phillip Sherrard, *The Sacred in Life and Art* (Ipswich, U.K.: Golgonooza Press, 1990), 71f.

82. Allen Carlson, "Appreciation and the Natural Environment," in *Environmental Ethics: Divergence and Convergence*, ed. Susan J. Armstrong and Richard G. Botzler (New York: McGraw-Hill, 1993), 142–48.

83. See Bruce V. Foltz, "Inhabitation and Orientation: Science Beyond Disenchantment," in *Earth Matters: The Earth Sciences, Philosophy, and the Claims of Community*, ed. Robert Frodeman (Upper Saddle River, N.J.: Prentice Hall, 2000), 26ff.

84. Leslie Marmon Silko, "Landscape, History, and the Pueblo Imagination," in *The Norton Book of Nature Writing*, ed. Robert Finch and John Elder (New York and London: Norton, 1990), 884f.

85. Ouspensky, 189.

86. St. Symeon the New Theologian, *On the Mystical Life: The Ethical Discourses*, Vol. I, trans. Alexander Golitzin (Crestwood, N.Y.: St. Vladimir's Seminary Press, 1995), 29.

87. Olivier Clément, *Conversations with Ecumenical Patriarch Bartholomew I* (Crestwood, N.Y.: St. Vladimir's Seminary Press, 1997), 105, 102.

88. Evdokimov, 2, 26.

89. Konstantin Mochulsky, *Dostoevsky: His Life and Work*, trans. Michael A. Minihan (Princeton, N.J.: Princeton University Press, 1971), 380.

90. John Muir, like many nature writers, saw civilization itself as Fall (in his words, as "encrusting dirty sin"), and thus felt that it was because the will of God was still discernable in wilderness, that humanity stood in desperate need of its message. See Dennis Williams, "John Muir, Christian Mysticism, and the Spiritual Value of Nature," 96, in *John Muir: Life and Work*, Sally M. Miller ed., (Albuquerque: University of New Mexico Press, 1993). Williams makes a strong case here that Muir should be regarded as one of the great Chrisitan mystics.

91. Edward Abbey, *Desert Solitaire: A Season in the Wilderness* (New York: Ballantine Books, 1971), 189f. Abbey adds that Paradise for him would need to have Gila monsters as well as cherubim.

92. Abbey, 191f.

93. Julia Kristeva, *In the Beginning Was Love: Psychoanalysis and Faith* (New York: Columbia University Press, 1987), 23.

94. In *Devils*, and in each of the following works, all three of these ills are linked to Western "scientific" thought.

95. Fyodor Dostoevsky, *Demons*, trans. Richard Pevear and Larissa Volokhonsky (New York: Vintage Books, 1995), 255.

96. Fyodor Dostoevsky, *A Writer's Diary*, Vol. 1, 1873–1876, trans. Kenneth Lantz (Evanston: Northwestern University Press, 1993), 591f.

97. Mikhail Bakhtin, *Problems of Dostoevsky's Poetics*, ed. and trans. Caryl Emerson (Minneapolis: University of Minnesota Press, 1984), 150.

98. Fyodor Dostoevsky, *A Writer's Diary*, Vol. 2, 1877–1881, trans. Kenneth Lantz (Evanston: Northwestern University Press, 1994), 952ff.

99. The heart of this teaching can be found in "Book Six: The Russian Monk."

100. Fyodor Dostoevsky, *The Brothers Karamazov*, trans. Richard Pevear and Larissa Volokhonsky (New York: Vintage Books, 1991), 288f.

101. Dostoevsky, *Karamazov*, 299.

102. Dostoevsky, *Karamazov*, 319f.

103. Pavel Florensky, *The Pillar and Ground of the Truth* (Princeton: Princeton University Press, 1997), 192.

104. Elder Ephraim of the Holy Monastery of St. Anthony, "On Salvation and Paradise," *Divine Ascent*, Number 5, Autumn, 1999, 103.

105. That this vision is by no means geographically fixed is indicated by its continuity with the spiritual and aesthetic vision of Celtic Christianity, which to some extent survived in the West, due largely to its remoteness from Western centers of power.

106. Seyyed Hossein Nasr, *Religion and the Order of Nature* (New York: Oxford University Press, 1996), 7.

107. Philip Sherrard, *Christianity: Lineaments of a Sacred Tradition* (Brookline, Mass.: Holy Cross Orthodox Press, 1998), 214.

108. Dostoevsky, *Karamazov*, 320.

109. John Muir, *The Story of My Boyhood and Youth* (Boston and New York: Houghton and Mifflin, 1926), 228; italics added. See note 1 of this chapter.

110. Patrick D. Murphy, "Penance or Perception: Spirituality and Land in the Poetry of Gary Snyder," in John Cooley, *Earthly Words*, 242ff.

111. Nasr, 29–66.

112. Massimo Cacciari, "Porte della vita"/"Life-gates," in *Figurare L'Invisible: Icone greche della Collezione Velimezis a Venezia* (Milan: Skira editione, 2000), 21f.

7. Seeing Nature: *Theōria Physikē* in the Thought of St. Maximos the Confessor

1. p. 320.

2. Maximos the Confessor, *Mystagogy*, in Hans Urs von Balthasar, *Cosmic Liturgy: The Universe According to Maximos the Confessor* (San Francisco, Ignatius Press, 2003), 327.

3. "Hans-Georg Gadamer, "Heidegger's Later Philosophy," in Gadamer, *Philosophical Hermeneutics*, trans. David E. Ling (Berkeley: University of California Press, 1976), 216f.

4. Friedrich Nietzsche, *Thus Spoke Zarathustra*, in *The Portable Nietzsche*, ed. and trans. Walter Kaufmann (New York: Viking, 1954), 125.

5. Ibid.

6. Friedrich Nietzsche, *Beyond Good and Evil*, in *Basic Writings of Nietzsche*, ed. and trans. Walter Kaufmann (New York: Modern Library, 1968), 265.

7. Martin Heidegger, "The Will to Power as Knowledge and as Metaphysics," in *Nietzsche: Volumes Three and Four* (San Francisco: Harper & Row, 1991), Volume III, pp. 17, 231f, 240.

8. Ibid, 245.

9. Marcel Gauchet, *The Disenchantment of the World: A Political History of Religion*, trans. Oscar Burge (Princeton: Princeton University Press, 1997), 95.

10. Plutarch, *On the Disappearance of Oracles*, XIV, 417d, cited in Pavel Florensky, *The Pillar and Ground of Truth*, trans. Boris Jakim (Princeton: Princeton University Press, 1997), 529.

11. Documenting this re-sacralization, historian Peter Brown describes how through holy relics—and much more importantly in the East, the *life* and *being* of holy men and women, monastics and saints and holy fools—"paradise itself came to ooze into the world. Nature itself was redeemed. . . . The countryside found its voice again . . . in an ancient and spiritual vernacular, of the presence of the saints. Water became holy again. The hoofprint of his donkey could be seen beside a healing spring, which St. Martin had caused to gush forth from the earth . . . They brought down from heaven to earth a touch of the unshackled, vegetable energy of God's own paradise." Peter Brown, *The Rise of Western Christianity: Triumph and Diversity, A. D. 200–1000* (Oxford: Blackwell, 2003), 164.

12. Pavel Florensky, *The Pillar and Ground of Truth: An Essay in Orthodox Theodicy in Twelve Letters*, trans. Boris Jakim (Princeton: Princeton University Press, 1997), 210.

13. St. Isaac of Syria, *Daily Readings with St. Isaac of Syria*, ed. A. M. Allchin, trans. Sebastian Brock (Springfield, Ill.: Templegate, 1989), 29.

14. Peter Kalkavage, trans., *Plato's Timeaus* (Newburyport, Mass.: Focus Publishing, 2001), 128.

15. Andrew Louth, *The Origins of the Christian Mystical Tradition* (Oxford: Oxford University Press, 1981), 20.

16. Ibid., 53.

17. *Commentary on the Song*, III. 12, cited in Louth, *Origins*, 59.

18. Gregory of Nyssa, *The Life of Moses*, trans. Everett Malherve and Everett Ferguson (New York: Paulist Press, 1978), 137, 96.

19. Hans Urs van Balthasar, *Presence and Thought: An Essay on the Religious Philosophy of Gregory of Nyssa* (San Francisco: Ignatius Press, 1995), 48f, 92f.

20. Evagrios the Solitary, "Texts on Discrimination in Respect of Passion and Thoughts," in *The Philokalia*, Volume 1 (London: Faber and Faber, 1979), 42f.

21. Evagrios the Solitary, "On Prayer," in *The Philokalia*, Volume 1, 61f. The translation here also interpolates that of John McGuckin in *The Book of Mystical Chapters: Meditation on the Soul's Ascent from the Desert Fathers and Other Early Christian Contemplatives* (Boston: Shambala, 2002), 83; italics added.

22. The relation between kataphatic (positive) and apophatic (negative) theology in Dionysios is put succinctly by Eric Perl: "To pass from the intellectual apprehension of being to the 'mystical' encounter with the One is, once again, like turning from a multiplicity of reflections to that which is being reflected." Eric D. Perl, *Theophany: The Neoplatonic Philosophy of Dionysios the Areopagite* (Albany: SUNY Press, 2007), 94.

23. Hans Urs von Balthasar, *The Glory of the Lord: A Theological Aesthetics*, Volume II, trans. Andrew Louth, Francis McDonagh, and Brian McNeil (San Francisco: Ignatius Press, 1984), 182.

24. David Bradshaw, *Aristotle East and West: Metaphysics and the Division of Christendom* (Cambridge: Cambridge University Press, 2004), 206.

25. Ibid.

26. Maximos Confessor, *The Church's Mystagogy*, in *Selected Writings*, trans. George C. Berthold (New York: Paulist Press, 1985), 189.

27. St. Maximos the Confessor, *Third Century of Various Texts*, in *The Philokalia*, Volume 2, ed. St. Nikodimos of the Holy Mountain and St. Makarios of Corinth, trans. G. E. H. Palmer, Philip Sherrard, Kallistos Ware (London: Faber and Faber, 1981), 121.

28. Maximos the Confessor, *Difficulty [Ambiguum] 41*, in *Maximos the Confessor*, ed. and trans. Andrew Louth (London and New York: Routledge, 1996), 158.

29. Steven Cassedy, *Dostoevsky's Religion* (Stanford: Stanford University Press, 2005), xiii, 162, 175

30. Wendell Berry, *A Continuous Harmony: Essays Cultural and Agricultural* (New York: Harcourt Brace Jovanovich, 1970), 5f.

31. Thomas Merton, *The Hidden Ground of Love: The Letters of Thomas Merton* (New York: Farrar, Straus, Giroux, 1985), 50.

32. Thomas Merton, *The Inner Experience: Notes on Contemplation*, ed. William H. Shannon (San Francisco: Harper Collins, 2003), 68.

33. M. Basil Pennington, *Thomas Merton My Brother* (Hyde Park, N.Y.: New City Press, 1996), 110. See also pp. 26–28 and 98–113 for discussions of the influence of Byzantine spirituality on Merton.

8. Seeing God in All Things: Nature and Divinity in Maximos, Florensky, and Ibn 'Arabi

1. Basho, *The Narrow Road to the Deep North and Other Travel Sketches* (London: Penguin Books, 1966), 33.

2. Martin Lings, *A Sufi Saint of the Twentieth Century* (Cambridge: The Islamic Text Society, 1993), 136.

3. Martin Heidegger, *Elucidations of Hölderlin's Poetry*, trans. Keith Hoeller (Amherst, N.Y.: Humanity Books, 2000), 69.

4. Martin Heidegger, *Zollikon Seminars: Protocols—Conversations—Lectures*, ed. Medard Boss, ed. and trans. Franz Mayr and Richard Askay (Evanston, Ill.: Northwestern University Press, 2001), 99.

5. Ibid.

6. Martin Heidegger, *Contributions to Philosophy (from Enowning)*, trans. Parvis Emad and Kenneth Maly (Bloomington: Indiana University Press, 1999), 195.

7. Take, for example, the discourse of *die Göttliche*, gods or "divine ones." Sometimes this seems to characterize only early Greek culture, before the inception of philosophy, while at the same time it somehow remains exemplary for us today. Sometimes the discourse of "divine ones" simply seems to designate one revelatory tradition among others: the coming into presence "of the divine in the world of the Greeks, in prophetic Judaism, in the preaching of Jesus."[7] In some essays, however, *die Göttlichen* seem to constitute a necessary dimension of the fourfold structure of *any* world. And in the *Beitrage*, what seems to be important is not divine ones or gods at all, but the hidden sway of a final God (*letzte Gott*).

8. Many of the difficulties are due to the historicism peculiar to Heidegger's thought, i.e., his view that experience is not only shaped and contextualized by its respective historical epoch, but that it is thoroughly determined by it. But this is not a vulgar historicism. Heidegger knows well enough—and indeed, he helped contemporary thought to understand—that experience itself is based upon what is not experienced, and its explication based upon what remains unsaid. For Heidegger, however, even that which is not exhausted within experience itself—that residue of experience upon which experience itself depends—that which remains unsaid and unsayable in the saying, is *itself* historically determined: as being nothing more than that which *for a given epoch* remains the unexperienced, the unsaid, the unthought. The unsayable and unthought and unexperienced *for us* would thus be nothing more than *our* unsayable and unthought and unexperienced, fixed within the constellation of being and thinking that we, as temporal beings, necessarily inhabit, and by no means *the same* unsayable and unthought and unexperienced that ancient Greeks or medieval Byzantines or nineteenth century Germans sought to say, think, and apprehend. Apophaticism here gets re-aligned toward a moving target.

9. James, *Varieties of Religious Experience*, 328, 370.

10. Maximos Confessor, *The Church's Mystagogy*, in *Selected Writings*, translated by George C. Berthold (New York: Paulist Press, 1985), 189, and interpolated with the translation of the same passage in Olivier Clément, *The Roots of Christian Mysticism* (London: New City Press, 1993), 219.

11. Amb. 42, 85, also in *The Philokalia*, Volume 2, ed. St. Nikodimos of the Holy Mountain and St. Makarios of Corinth, trans. G. E. H. Palmer, Philip Sherrard, and Kallistos Ware (London: Faber & Faber, 1981), 268.
12. Ad Thal. 2, 100.
13. Ibid., 101.
14. *Philokalia*, Volume 2, 124.
15. Muhyîddîn Ibn 'Arabi, *Futûhât al-Makkiyya*, 319, cited in Stephen Hirtenstein, *The Unlimited Mercifier: The Spiritual Life and Thought of Ibn 'Arabi* (Oxford: Anqa Publishing), 1999, 102.
16. *Futûhât*. II, 684, SDG 91.
17. *Futûhât*. II, 423, cited in *Unlimited Mercifier*, 98f.
18. *K. al-'Abadilah*, 42, cited in *Unlimited Mercifier*, 28.
19. Henry Corbin, *Alone with the Alone: Creative Imagination in the Sufism of Ibn 'Arabi* (Princeton: Princeton University Press, 1969), 145.
20. *Futûhât*. II, 324, cited in *Alone*, 146.
21. *Futûhât*. II, 326, cited in *Alone*, 330.
22. *Futûhât*. II, 331, cited in *Alone*, 332.
23. Pavel Florensky, *The Pillar and Ground of Truth: An Essay in Orthodox Theodicy in Twelve Letters*, trans. Boris Jakim (Princeton: Princeton University Press, 1997), 212f.
24. Ibid., 229.
25. Ibid., 235, 283.
26. Ibid., 236f.

9. The Glory of God Hidden in Creation: Eastern Views of Nature in Fyodor Dostoevsky and St. Isaac the Syrian

1. Cited in Olivier Clément, *The Roots of Christian Mysticism* (London: New City Press, 1993), 213.
2. St. Isaac of Ninevah, *[The Ascetical Homilies:] 'The Second Part,' Chapters IV–XLI, Scriptores Syri, Tomus 225*, trans. Sebastion Brock (Lovani [Belgium]: Aedibus Peeters, 1995), 160.
3. Lynn White Jr., "The Historical Roots of Our Ecologic Crisis," in *Dynamo and Virgin Reconsidered: Essays in the Dynamism of Western Culture* (Cambridge, Mass.: MIT Press, 1971), 84.
4. Ibid., 93.
5. Ibid., 90.
6. Ibid., 93.
7. St. Isaac the Syrian, Homily 77, *Ascetical Homilies*, cited in Hilarion Alfeyev, *The Spiritual World of St Isaac the Syrian*, Cistercian Series Number One Hundred Seventy-Five (Kalamzaoo, Mich.: Cistercian Publications, 200), 114.
8. White, "Historical Roots," 88; italics added.
9. Steven Cassedy, *Dostoevsky's Religion* (Stanford, Calif.: Stanford University Press, 2005), 156.

10. Ibid., 159.

11. Ibid., xii, 162, 175.

12. *Dostoevsky Studies*, vol. VII, 1986, 142.

13. Stella Rock's important work, *Popular Religion in Russia: "Double Belief" and the Making of an Academic Myth* (*Routledge Studies in the History of Russia and Eastern Europe*) (New York: Routledge, 2007) presents an authoritative refutation of the claim that Russian popular religion exhibits, in any extraordinary sense, a "double faith" in both Christian and pagan beliefs and practices. To the contrary, the "two faiths" signified by the old Russian word *dvoeverie* had always been Roman Catholicism and Eastern Orthodoxy, with the revisionist meaning largely put into play by the Soviets. Rock is also critical of the tendency of Western commentators to posit a very narrow, self-assured view of what they consider to be "Christian" and then use this as a standard for judging the legitimacy of Russian Orthodoxy.

14. St. Isaac of Syria, *Daily Readings with St. Isaac of* Syria, ed. A. M. Allchin, trans. Sebastian Brock (Springfield, Ill.: Templegate, 1989), 29.

15. See, for example, Victor Terras, *A Karamozov Companion: Commentary on the Genesis, Language, and Style of Dostoevsky's Novel* (Madison: University of Wisconsin Press, 2002), 22ff; Rowan Williams, *Dostoevsky: Language, Faith, and Fiction* (Waco, Tex.: Baylor University Press, 2008), 204f.

16. Nicholas Berdyaev, *Dostoevsky*, trans. Donald Attwater (Cleveland and New York: Meridian Books, 1965), 62ff.

17. St. Isaac the Syrian, *The Ascetical Homilies of St Isaac the Syrian*, trans. Holy Transfiguration Monastery (Boston: Holy Transfiguration Monastery, 1984), 229.

18. Fyodor Dostoevsky, *The Adolescent*, trans. Richard Pevear and Larissa Volokhonsky (New York: Vintage, 2003), 355.

19. Photios Kontoglou, "Econium: An Offering of Praise to Saint Abba Isaac the Syrian, Inadequate for the Sublimity of Its Subject, but Written with Much Love," in *The Ascetical Homilies of St Isaac the Syrian*, 2nd edition (Boston: Holy Transfiguration Monastery, 2011), 49.

20. Fyodor Dostoevsky, *The Brothers Karamozov*, trans. Richard Pevear and Larissa Volokhonsky (New York: Vintage, 1991), 320.

21. *Ascetical Homilies (1984)*, 41.

22. Ibid., 300.

23. St. Isaac of Ninevah, *[The Ascetical Homilies:] Second Part*, 14.

24. Dostoevsky, *Karamaov*, 319.

25. St. Isaac of Ninevah, *[The Ascetical Homilies:] Second Part*, 151; italics added. St. Isaac is citing here from Psalm 103/104:15.

26. *The Festal Menaion*, trans. Mother Mary and Archimandrite Kallistos Ware (South Canaan, Pa.: St Tikhon's Seminary Press, 1998), 354–62.

27. For a suggestive collection of articles comparing the Orthodox and Sufi views of the heart and its central role in the spiritual life, see James Cutsinger (ed.), *Paths to the Heart: Sufism and the Christian East* (Bloomington, Ind.: World Wisdom, 2004).

28. After a trip to the Taos Pueblo, the depth psychologist Carl Jung reported: "Our theory is that the seat of consciousness is in the head, but the Pueblo Indians told me that Americans were mad because they believed their thoughts were in their heads, whereas any sensible man knows that he thinks with his heart." C. G. Jung, *Modern Man in Search of a Soul* (New York: Harcourt, 19330, 184.

29. "Though I do not believe in the order of things, still the sticky little leaves that come out in the spring are dear to me, the blue sky is dear to me . . . Sticky spring leaves, the blue sky—I love them, that's all!" *Karamazov*, 230.

30. *The Philokalia*, Volume 4, ed. St. Nikodimos of the Holy Mountain and St. Makarios of Corinth, trans. G. E. H. Palmer, Philip Sherrard, Kallistos Ware (London: Faber and Faber, 1995), 93.

31. *The Philokalia*, Volume 1, ed. St. Nikodimos of the Holy Mountain and St. Makarios of Corinth, trans. G. E. H. Palmer, Philip Sherrard, Kallistos Ware (London: Faber and Faber, 1979), 194.

32. Dostoevsky, *Karamazov*, 289; italics added.

33. St. Isaac of Ninevah, *[The Ascetical Homilies:] Second Part*, 151ff; *Ascetical Homilies*, "Homily Sixty-Five," 322; italics added.

34. Ivan's atheism, his "blasphemy," is directed not primarily against God, but against the world understood as divine creation. What is the blasphemy here? Not the assertion that there is no God, but that creation is meaningless: "The theme of [these two sections] is blasphemy and the refutation of blasphemy . . . As it manifests itself in Russia among (almost) our entire upper crust, and principally among the younger generation, i.e., the scientific and philosophical refutation of the existence of God having been already discarded, the present day serious socialists no longer bother with it. . . . Instead, they vehemently deny God's creation, God's world and its significance." (Letter of May 19, 1979 in *Selected Letters of Fyodor Dosoyevsky*, ed. Joseph Frank and David Goldstein, trans. Andrew MacAndrew (New Brunswick, N.J.: Rutgers University Press, 1989). Despite his admission that it is principally nature that somehow allows him to keep on living, Ivan is nevertheless insistent that creation is pointless and cruel. Indeed, it is this meaninglessness that makes the Christ of Ivan's "Legend" an impotent figure because He is an anomaly in the real order of things. Ivan, and the Inquisitor, are proto-existentialists, proclaiming the meaninglessness and absurdity of creation—a meaninglessness that grants a license to all those Promethean rationalists who would undertake to re-design the created order.

35. Dostoevsky, *Karamazov*, 362.

10. Between Heaven and Earth: Did Christianity Cause Global Warming?

1. Annie Dillard, "Feast Days: Thanksgiving—Christmas III," in *Upholding Mystery: An Anthology of Contemporary Christian Poetry*, ed. David Impastato (Oxford: Oxford University Press, 1997), 148.

2. Martin Heidegger, "The Question Concerning Technology," in *The Question Concerning Technology and Other Essays*, trans. William Lovitt (New York: Harper & Row, 1977), 14.

3. Martin Heidegger, "Memorial Address" [*Gelassenheit*], in *Discourse on Thinking*, ed. Martin Heidegger, trans. John M. Anderson and E. Hans Freund (New York: Harper and Row, 1966), 50.

4. Lynn White, Jr., *Medieval Technology and Social Change* (Oxford: Oxford University Press, 1964), 79.

5. Ibid., 129.

6. Ibid., 134.

7. Ibid., 78, italics added.

8. For a fine discussion of the development of the philosophical and theological concept of *energeia* in the Byzantine East, see David Bradshaw, *Aristotle East and West: Metaphysics and the Division of Christendom* (Cambridge: Cambridge University Press, 2007). He discusses the Cappadocians at some length on pages 172–78.

9. Michael Allen Gillespie, *Nihilism Before Nietzsche* (Chicago: University of Chicago Press, 1995).

11. Nature and Other Modern Idolatries: *Kosmos*, *Ktisis*, and Chaos in Environmental Philosophy

1. Christos Yannaras, *Variations on the Song of Songs*, trans. Norman Russell (Brookline, Mass.: Holy Cross Orthodox Press, 2005), 87.

2. Maurice Merleau-Ponty, *Nature: Course Notes from the Collège de France*, ed. Dominique Séglard, trans. Robert Vallier (Evanston, Ill.: Northwestern University Press, 2003), 85.

3. Kathleen Raine, *Blake and Antiquity* (London: Routledge, 2002), 100; William Blake, "Augeries of Innocence."

4. Indeed, the Latin words *natura* and *essentia* had been linked since late antiquity, an inheritance from Greek philosophy for which the corresponding terms *physis* and *ousia* had been similarly identified.

5. It is notable that in ancient Greek philosophy, on those few occasions where *physis* was used to designate "the all" of what is, it included not just cosmology, but human affairs as well, such as politics. See Gerard Naddaf, *The Greek Concept of Nature*, trans. Gerard Naddaf, SUNY Press, 2005), 35. And even in Roman philosophy, Lucretius's great materialist poem about the cosmos bears the title *De Rerum Natura*, where *natura* refers to what things are, what they are like, rather than to physical things themselves taken together.

6. Martin Heidegger, *Zollikon Seminars: Protocols—Conversations—Lectures*, ed. Medard Boss, trans. Franz Mayr and Richard Askay (Evanston, Ill.: Northwestern University Press, 2001), 26.

7. Owen Barfield, *History in English Words* (Barrington, Mass.: Lindisfarne, 1976), 170f.

8. On the concept of the "man-God," see Chapter 9; for Heidegger's most sustained elaboration of his critique of modern humanism, see his "Letter on Humanism."

9. Michael Allen Gillespie, *The Theological Origins of Modernity* (Chicago: University of Chicago Press, 2009), 274; italics added.

10. Alexandre Koyré, *From the Closed World to the Infinite Universe* (Baltimore: Johns Hopkins University Press, 1968), 1f.

11. Ibid., 61.

12. William Blake, "The Marriage of Heaven and Hell."

13. Raine, 99.

14. S. Foster Damon, *A Blake Dictionary: The Ideas and Symbols of William Blake* (Lebanon, N.H.: University Press of New England, 1988), 76f.

15. Thomas Kuhn, *The Copernican Revolution*, cited in Dennis Danielson, *The Book of the Cosmos: Imagining the Universe from Heraclitus to Hawking* (Cambridge, Mass.: Basic Books, 2000), 120f.

16. David Hume, *The Natural History of Religion*, Section XV, "General Corollary."

17. Remi Brague, *The Wisdom of the World: The Human Experience of the Universe in Western Thought*, trans. Teresa Lavender Fagan (Chicago: University of Chicago Press, 2003), 19.

18. Margaret Barker, *Creation*, 37.

19. Hans Urs von Balthasar, *The Glory of the Lord: Volume I*, trans. Erasmo Leiva-Merikakis (San Francisco: Ignatius Press, 1982), 431.

20. The exhortation to look away from the external world in order to find God within the soul is often attributed to Augustine. There are certainly strong textual grounds for this claim, but at the same time Augustine is by no means impervious to the possibility of finding at least evidence of divinity—if not a sense of divine presence—in the beauty of creation.

21. I am indepted to John Behr's discussion of Athanasios's idolatry interpretation in *The Nicene Faith, Part One: True God of True God* (Crestwood, N.Y.: St. Vladimir's Seminary Press, 2004), 168–84, for drawing my attention to its importance, as well as to Kyle Hofmeister for pointing me to this text.

22. *The Essence of Christianity*, 13.

23. Ibid., 116.

24. Jean-Luc Marion, *The Idol and Distance: Five Studies*, trans. Thomas A. Carlson (New York: Fordham University Press, 2001), 7.

25. David Bentley Hart, *The Beauty of the Infinite: The Aesthetics of Christian Truth* (Grand Rapids, Mich.: Eerdmans, 2003), 269.

26. Emmanuel Levinas, *Difficult Freedom: Essays on Judaism* (Baltimore: The Johns Hopkins University Press, 1990), 14. Levinas seems to ignore here the rich commentary (discussed earlier in this chapter) on the glory of God as manifest in creation to be found throughout the Psalms, in the concluding section of the Book of Job, and in the Wisdom Books in general, not to mention the encounter with God by Moses on Mt. Sinai.
27. Christos Yannaras, *Person and Eros*, trans. Norman Russell (Brookline, Mass.: Holy Cross Orthodox Press, 2007), 22f.
28. Ibid., 5.
29. Ibid., 18.

12. Traces of Divine Fragrance, Droplets of Divine Love: The Beauty of Visible Creation in Byzantine Thought and Spirituality

1. Pavel Florensky, *The Pillar and Ground of Truth: An Essay in Orthodox Theodicy in Twelve Letters*, trans. Boris Jakim (Princeton: Princeton University Press, 1997), 201.
2. Ibid., 210.
3. Ibid.
4. *The Philokalia: The Complete Text*, Volume 4, compiled by St. Nikodemos of the Holy Mountain and St. Makarios of Corinth, trans. G. E. H. Palmer, Philip Sherrard, and Kallistos Ware (London: Faber and Faber, 1995), 92.
5. *The Philokalia: The Complete Text*, Volume 1, compiled by St. Nikodemos of the Holy Mountain and St. Makarios of Corinth, trans. G. E. H. Palmer, Philip Sherrard, and Kallistos Ware (London: Faber and Faber, 1979), 194.
6. Sisters of the Holy Convent of Chrysopigi, *Wounded by Love: The Life and Wisdom of Elder Porphyrios* (Limni, Evia, Greece: Denise Harvey, 2005), 140.
7. Protocol of an anonymous pilgrim, cited in *Hieromonk* Isaac, *Elder Paisios of Mount Athos*, trans. *Hieromonk* Alexis (Trader) and Fr. Peter Heers, ed. *Hieromonk* Alexis (Trader), Fr. Evdokimos (Goranitis) and Philip Navarro (Chalkididi, GR: Holy Monastery "St Arsenios the Cappadocian," 2012), 238.
8. Florensky, *Pillar and Ground of Truth*, 200, 527.
9. Ibid., 216; italics added.
10. D. S. Wallace-Hadrill, *The Greek Patristic View of Nature* (Manchester: Manchester University Press, 1968), 87–91.
11. St. Basil the Great, Letter XIV, in *Nicene and Post-Nicene Fathers, Second Series*, Volume 8, ed. Philip Schaff (Peabody, Mass.: Hendrickson, 1999).
12. *Hexaemeron*, 71, 55.
13. Ibid., 76.

14. Gregory of Nyssa, Letter IX, in *Nicene and Post-Nicene Fathers, Second Series,* Volume 5, ed. Philip Schaff (Peabody, Mass.: Hendrickson, 1999), 532.

15. *The Natural Beauty of Mt. Athos* (Thessaloniki: Hagioritki Hestia, 2003), n.p.

16. Elder Ephriam, *Counsels from the Holy Mountain: Selected from the Letters and Homilies of Elder Ephraim* (Florence, Ariz.: St. Anthony's Greek Orthodox Monastery, 1998), 1; italics added.

17. Sisters of the Holy Convent of Chrysopigi, *Wounded by Love,* 81.

18. Elder Joseph the Hesychast, *Monastic Wisdom: The Letters of Elder Joseph the Hesychast* (Florence, Ariz.: St. Anthony's Greek Orthodox Monastery, 1998), 270.

19. *Prayers by the Lake,* 56.

20. Pilgrim, 77.

21. *The Philokalia: The Complete Text,* Volume 3, compiled by St. Nikodemos of the Holy Mountain and St. Makarios of Corinth, trans. G. E. H. Palmer, Philip Sherrard, and Kallistos Ware (London: Faber and Faber, 1984), 171.

22. Dostoevsky, *Karamozov,* 299.

23. Gervase Mathew, *Byzantine Aesthetics* (New York: Harper & Row, 1971), 1–6.

24. Ibid., 6.

25. St. Theodoros the Great Ascetic, *Theoretikon,* in *The Philokalia,* Volume 2, ed. St. Nikodimos of the Holy Mountain and St. Makarios of Corinth, trans. G. E. H. Palmer, Philip Sherrard, and Kallistos Ware (London: Faber & Faber, 1981), 44.

26. St. Gregory of Nyssa, *Homiliae I,* 22.

27. Ibid., 33, translation altered.

28. *Wisdom of Solomon,* 7:1–6.

29. *The Philokalia: The Complete Text,* Volume 2, comp. St. Nikodemos of the Holy Mountain and St. Makarios of Corinth, trans. G. E. H. Palmer, Philip Sherrard, and Kallistos Ware (London: Faber and Faber, 1981), 42–45.

Index of Terms in Greek, German, and Latin

Greek Terms and Phrases

agapē, 35
agathōn, 148
angelos, angelois, 69
aisthēsis, 4, 32, 144
to aisthetoi kallos, 242
anamnēsis, 108, 135, 261
anthrōpos, 160
askēsis, xvii, 131, 149, 163, 209

bios theōrētikos, 3

chōrismos, 14, 75, 167
chronos (tōn chronōn kai tōn kairōn), 59

daimonia, 235
Dēmiourgos, 220
diakosmēsis, 170
dianoia, 6, 129, 197, 198, 249
doxa, 5, 20, 59, 221, 222

eidos, eidē, 8, 9, 18
eikōn, x, 4, 9, 10, 126, 128, 168
eirēnē kai asphaleia, 60
ekstasis, 230
energeia, 11, 14, 96, 193, 204, 208, 211, 250, 275; *energeiai*, 129, 185, 209, 210
epictasis, 136
epikeina, 160
epiphaneia, 135
epochē, 93, 131
erōs, 35, 224, 230, 231, 242, 243

Hagia Sophia, 46. *See also* Hagia Sophia *in the main index*
haidēs, Haidēs, 3, 9
(to) Hen, 220
hesychia, 211
hypo-stasis, 229

to idiazon, 230
idios, 223

kairos (tōn chronōn kai tōn kairōn), 59
kardia, 4
kalon, 147, 164
katharsis, 11, 66, 72, 127, 167, 175, 179, 199, 239, 250
kenōsis, 231
klēsis, 230

komboskini, xv
kosmikoi, 86
kosmos, viii, xvii, 21, 59, 60, 161, 163, 164, 181, 182, 215, 216, 220, 221, 223, 224
kosmos aisthetos, 148
kosmos noētos, 148
ktisis, viii, xvii, 21, 215, 216, 220, 221, 222, 224

logismoi, 4, 197, 198, 250
logos, 12, 83, 106, 128, 129, 163, 169, 171, 180; *logoi*, 12, 72, 82, 85, 104, 104, 128, 130, 147, 166–71, 180–82, 184, 193, 194, 202, 209, 241, 260; *Logos*, 82, 83, 126, 128, 141, 181, 209, 216, 220 *See also* Logos *in the main index*
logos physeōs, 128
logos spermatikos, 104

metanoia, 198
methexis, 14, 17
mimēsis, 84
moira, 7
mythos, mythoi, 9, 144, 145

nēpsis, 19, 197, 243, 150
noēsis, 6, 8, 32, 33, 66
nous, 7–9, 129, 179, 197, 199, 201, 213, 223, 224, 236–39, 244, 248, 249; *Nous*, 46, 218, 220

oikos (ecos), 99, 100, 154
ontōs on, 9
orthos doxa, 197
ousia, 10, 11, 96, 129, 185, 193, 209, 210, 229, 275

parousia, 60
phainetai, 230
philokalia, 66, 96, 235. *For the collection of texts with this name, see the main index*
physikē, 166, 170
physis, 9, 10, 36, 40, 50, 55, 79, 87, 176, 177, 204, 215, 216, 219, 229, 230, 241, 275; *Physis*, 84
poiēsis, 78, 203
polemos, 132
polis, 76, 87
praxis, 95, 180
prosōpon, 221, 229–31; *pros-ōpon*, 229
pros ti, 230
prōtē anthrōpos, 244

Sophia, 216, 220. *See also* Sophia *in the main index*
symbolon, 136, 250; *syn-holon*, 33, 218, 250

taxis, 198
technē, 77, 78, 241, 242
thaumazein, 3, 242; *thaumazō*, 100
thea, 4
theasthai, 3
theologia, 118, 262
theologos, 118
theōrein, 166
theōria, 3, 4, 11, 12, 14, 20, 60, 62, 66, 67, 164, 174, 179, 180, 199, 199, 213, 239; *theōreitai, 165*
theōria physikē, vii, 12, 20, 67, 72, 128, 158, 165–74, 180, 199, 209, 211, 239, 241, 257
theōros, theōroi, 3, 4
theōsis, 11, 12, 72, 127, 170, 179, 199, 239
(ho) Theos, 220
to mē on, 9, 224
ta onta, 224
topos, 115

German Terms and Phrases

Abbau, 56
Aufhebung, 116

Begriff, 44
Besinnung, 4
Bestand, 55, 90, 99
Betrachtung, 3

Darstellung, 83, 135, 195
Dasein, 99, 230
Daseinsanalytic, 59
Denken, 4
Destruktion, 20, 56, 58, 60
Du, 45

Einsfühlung, 45
Entgotterung, 117
Entzauberung, 6, 161
Erde, 159; *irdisch*, 134; *Irdisch*, 159
Ereignis, 50, 62, 264
Erfahrung, 176

Faktizität, 57

Gefahr, 60
Gegebenheit, 50
Gegensatz, 56
Gegenstand, 55, 90; *Gegen-stand*, 48
Gestell, 36
Göttlichen, xiii, 62, 177, 271

Heil, 36
Heilen, 36
Heilige, 36
Heil-los, 36
Heilsgeschichte, 142
Herausfordern, 203
Holzwege, 210. *For the book of this name, see the main index*

Lebenswelt, 6
letzte Gott, 62, 118, 271

Pantheismusstreit, 14

Sache selbst, 26
Sagen, 118, 141
Schritt zurück, 99
Seinsgeschichte, 56, 62, 229
Stimmung, 141

Übermensch, 226
Unheil, 36

vorhanden, 3, 43, 52, 230; *Vorhandenheit*, 20
Verschlossenheit, 36
Verzauberung, 117
Vorstellung, 82, 135, 195

Weltanschauung, 3, 56

Zerstörung, 56
zuhanden, 50; *Zuhandenheit*, 101

Latin Terms and Phrases

actualitas, 204, 209 211, 250
actus, 209, 250
analogia entis, xvi, 14, 17, 83, 211
a posteriori, 212
axis mundi, 241

causa sui, 117
civitas, 180
cogito, 43
contemplatio, 3

de facto, 17
de jure, 17
destructio, 20

Deus, 216
deus absconditas, 59

ens creatum, 117, 121, 177, 188, 204, 213
essentia, 219, 275

factum, 14, 188

gratis, 214

mens, 7

natura, 219, 175; *Natura*, 216
natura naturans, xvi, 46, 104, 105, 109, 188
natura naturata, xvi, 104, 109

obiectum, 3

persona, 229
prima causa, 117

ratio, 218
rationes, 128
reductio ad absurdum, 121, 228
religio, 161
res cogitans, 43, 216
res extensa, 43, 216

speculum, 44
substantia, 229

theologia crucis, 58–60
theologia gloriae, 20, 58, 60

via moderna, 121, 126, 213

Index of Names and Places

Abbey, Edward, 115, 148, 149, 155
Abram, David, 45, 46, 48, 154, 222
Achaians, 220
Acropolis, 87
Adam, 9, 45, 86, 142, 189, 242, 244
Adams, Ansel, 104
Aegean Sea, 86, 126
Ajanta, 122
Al-Alawi, Shaikh Ahjad, 175, 181
Alaska, 71
Albion, 218
Alexandria, 63, 166, 208, 209, 222
Alps, 104; Bernese Alps, 233, Engadine Alps, 31, 94
America, 5, 31, 37, 148, 190; American Southeast, 104; American Southwest, x, 104; American West, 208; High Country of the American West, xiii
Americans, 274n28
American Transcendentalism, 14, 44, 94, 173
Ammonius Saccas, 166
Anatolia, 208
ancient Christianity, 192
ancient Greece, xiii, 5, 62, 233; ancient Greeks, xiii, 3, 6, 21, 45, 46, 75, 81, 161, 215, 235, 271n8
ancient Greek (language), 36, 248n12
ancient Jews, 21; Ancient Judaism, 21, 62, 63
Andalusia, 75, 178
Anselm, St., 13, 49, 211
Anthony the Great, St., 11, 163, 194, 199, 208, 236, 237
Antichrist, 60
Antioch, 86, 209
Apollo, 85, 161; the Apollonian, 32, 81
Apostle, 60
Aquinas, St. Thomas, 13, 15, 119, 122, 124, 210
Aristotle, 3, 8, 9, 14, 20, 29, 55, 166, 204, 208, 233, 242, 249n43, 262n3; *Generation of Animals*, 249n43
Artaud, Antonin, 227
Asia, 75, 76
Ataturk, Kemal, 95
Athanasios, St., xvi, 82, 222, 223, 224, 225, 226, 227, 228, 230, 242, 276n21; *Against the Hellenes*, 222

Athena, 87
Athens, 3, 8, 61, 62, 87, 164, 165, 168, 210, 222
Athonite, 126; Athonite Fathers, 238
Atlantic Ocean, xv
Attica, 76
Augustine, St., 14, 57, 61, 73, 128, 169, 173, 179, 185, 206, 208, 211, 230, 276n20
Aztecs, 81

Baath Party, 95
Bacon, Francis, 2, 13, 44, 52, 90, 218, 226
Baghdad, 179
Basil the Great, St., 208, 209, 235, 237; *Hexaemeron*, 209, 237
Bakhtin, Mikhail, 151
Bakunin, Mikhail, 95
Balkans, 73
Balthasar, Hans Urs von, 168, 222
Baptism of Christ, 196
Barker, Margaret, 221
Barlaam of Calabria, 129
Barr, James, 6
Bartholomew, Ecumenical Patriarch, ix, x, xiv, 265n49
Bartram, 31, 104
Basho, 175, 177, 178
Beatitudes, 214
Beauty, 28, 38, 125, 235
Becker. Oskar, 59
Behr, John, 276n21; *The Nicene Faith*, 276n21
Being, 19, 66, 141, 222
Being-in-the-World, 99
Belgic Confession, 70
Bering Sea, 69
Berleant, Arnold, 262n4
Berdyaev, Nikolai, 95, 98, 192
Bernard of Clairvaux, St., 61
Berry, Wendell, xii, 6, 14, 21, 49, 63, 71, 140, 151, 173, 174, 177, 220
Bethlehem, 84
Bible, 67, 158; *Deuteronomy*, 242; *Exodus*, 222; *Galatians*, 59; *Genesis*, xv, 74, 107, 120, 147, 149, 206, 213, 245; *Gospel of St. John* ("Prologue"), 166; *Isaiah*, 221; *Job*, 277n26; *1 John*, 245; *2 Peter*, 245; Prophetic Books, 66; Psalms, 46, 49, 63, 66, 67, 131, 165, 167, 221, 222, 277n26; *Proverbs*, 106; *Psalms*, 49, 235; *Revelation of St John*, 213, 221; *Romans*, 49; Septuagint Bible, 105, 106, 164, 216, 221, 235; *Song of Songs*, 243; *Thessalonians*, 59, 60; Wisdom Books, 59, 66, 67, 105, 243, 277n26; *Wisdom of Solomon*, 49, 59, 106, 165, 243
Bishop, Jeffrey P.: *The Anticipatory Corpse*, 254n26
Black Forest, 37
Blake, William, xii, 39, 41, 66, 97, 107, 110, 155, 173, 176, 199, 208, 209, 213, 215, 218, 219, 223, 227, 229; "Jerusalem," 39; *The Marriage of Heaven and Hell*, 179; "Milton," 215
Blavatsky, Helena, 95
Blessing of the Jordan River, 196
Blessing of the waters, 196
Boccaccio, Giovanni, 123
Bodhisattva, 108
Body, 45
Boehme, Jakob, 34, 103, 155
Bolsheviks, 95, 98
Bonaventure, St., 173
Bonaventure Cemetery, 68
Borneo, 9
Bosphorus Straits, 75, 76
Bradshaw, David, 169, 275; *Aristotle East and West*, 275n8

Brague, Remi, 220
Brann, Eva: *The Logos of Heraclitus*, 7
British Empiricism, 15
British Isles, 75
British Romanticism, 13, 44
Brown, Peter, 10, 85, 86, 161, 269n11; *The Rise of Western Christianity*, 269n11
Buber, Martin, 46
Buddha, 122
Buddhism, 82, 91, 108, 156, 234
Buddhist, 91, 108
Bulgakov, Sergei, 92, 95, 97–112, 261n47; *Capitalism and Agriculture*, 98; *The Philosophy of Economy*, 98, 99, 109, 110; *Vekhi (Landmarks)*, 98
Burroughs, John, xii, 40, 63, 94, 115, 144, 173
Byzantine: Christendom, 125; Christianity, 10, 34, 120, 121, 188, 189, 240, 242, East; xiii, 46, 61, 63, 66, 73, 86, 123, 124, 126, 172, 188, 206, 213, 241, 275n8; Empire, 122; Greek (language), 4, 248n12; Studies, 80
Byzantines, 271n8
Byzantium, xv, 74, 76, 77, 78, 79, 80, 81, 83, 86, 87, 96, 119, 121, 128, 155

Cacciari, Massimo, 156
Cairo, 179
California, 204
Calvin, John, 213
Cambridge Platonists, 70
Camus, Albert, 24, 115; "Helen's Exile," 24
Cappadocia, 10, 11, 63, 72, 208; Cappadocian Highlands, 166
Cappadocians, 193, 208, 209
Caputo, John, 16, 57, 119
Carlson, Alan, 143, 144
Carlyle, Thomas, 5; *Signs of the Times*, 5
Carolingian Renaissance, 155
Carson, Rachel, xii, 52
Caseau, Beatrice: *Guide to the Postclassical World*, 85
Cassidy, Steven, 190; *Dostoevsky's Religion*, 190
Caucasus Mountains, 109, 162
Catholicism, 57
Cave Monasteries of Kiev, 86
Chaos, 7, 134, 176, 218
Char, René, 24
Charlemagne, 13, 82, 83
Chateaubriand, François-René de, xii, 104
Chekhov, Anton, 95
Christ, 9, 12, 34, 65, 129, 135, 136, 140, 142, 196, 227, 232, 235, 238, 250–51n31, 265n49
Christendom, 55, 61, 121; Eastern, 80, 105; Latin, 125; Western, 212
Christian(s), 58, 60, 86, 116, 190, 216, 233, 239, 255n5; early, 21
Christian: East, 9, 11, 81, 85, 95, 105, 146, 163, 188, 193, 208; Middle Ages, 204; West, 73
Christianity, 20, 55, 57, 59, 60–63, 67, 68, 71, 72, 74, 81, 82, 91, 119, 120, 132, 153, 159, 161–63, 165, 166, 172, 177, 185, 187–92, 195, 196, 201, 202, 204–7, 209, 213, 214, 224, 233, 235–37, 240–42, 244; Celtic, 12, 268n105; Early, 9; Eastern, 38, 121, 131, 195, 206, 209, 210; Latin, 120, 185, 200; Western, 86, 154, 172, 191
Church, 12, 74, 121, 158, 200, 202; Eastern, 34, 111, 171, 262n3; Eastern Orthodox, 122; Roman Catholic, 122; Russian Orthodox, 96, 98; Western, 131, 185, 188

Cimabue, 123, 135
Clément, Olivier, 147
Cleopas, 167
Cohen, Michael, xi, 69, 247n7; *The Pathless Way: John Muir and the American Wilderness*, 247n7
Coleridge, Samuel Taylor, xii, 28; *Anima Poetae*, 28
Confucius, 107
Constantine, 76
Constantinople, ix, xiii, 1, 63, 75, 76, 78, 79, 81, 105, 122, 125, 127, 166, 178, 209; New Rome, 76, 79, 122, 155
Continental Rationalism, 15
Copernicus, 104, 220
Corbin, Henri, 182, 183
Cosmic Liturgy, 169
Cosmos, 218
Creation, 45, 221
Creator, xvi, 12, 13, 15, 20, 37, 58, 59, 66, 68, 130, 148, 153, 183, 195, 196, 208, 209, 233, 238, 242–45, 256n25, 265n49
Crichton, Michael, 89, 90, 91
Crowe, Benjamin D., 56
Crystal Palace, the, 200
Cutsinger, James: *Paths to the Heart: Sufism and the Christian East*, 274n27

Damascus, 179
Dante, 123; *Purgatorio*, 123
Dark Ages, 80, 121, 192, 211
Darwin, Charles, 206
Day of the Lord, 59, 60
Deist God, 213
Deleuze, Gilles, 16, 17, 224, 226, 227, 228, 229
Derrida, Jacques, xiii, 16, 17, 50, 75, 93, 119, 210
Descartes, René, 13, 16, 24, 26, 43, 44, 119, 176, 207, 211–13, 216, 218, 256n25; *Discourse on Method*, 207
Desert Fathers, 72, 86, 184
Desert Mothers, 72, 86, 184
Devil, 125, 198
Dillard, Annie, xii, 14, 40, 48, 50, 63, 67, 94, 104, 107, 115, 131, 136, 137–39, 141, 144, 155, 173, 174, 177, 180, 199, 203, 220; *Pilgrim at Tinker Creek*, 107, 136
Diné, x
Dionysian, 32, 81
Dionysius the Areopagite, St., 12, 58, 103, 168, 210, 234, 243, 270n22; *Mystical Theology*, 129, 168; *On the Celestial Hierarchy*, 168, 235; *On the Divine Names*, 168, 235
Dionysus, 85, 226, 235
Divine Countenance, 39
Divine Goodness, 235
Divine Liturgy, 10,12, 84, 169, 170, 240
Divine Nature, 245
Divine Wisdom, 34, 38, 83, 105, 106, 183, 185, 201
Divine Word, 166
Divinity, 162, 185, 236
Donne, John, 52
Dostoevsky, Fyodor, 33, 61, 62, 68, 73, 87, 94–96, 106, 148, 149, 150–52, 163, 172, 188–201, 217; Alyosha Karamazov (*Brothers Karamozov*), 113, 150, 190, 199, 202; Arkady Svidrigailov *(Crime and Punishment)*, 33; *The Brothers Karamozov*, 73, 113, 150, 151, 158, 189, 190, 194, 199–201, 239; *Crime and Punishment*, 33, 189, 201; *Devils (Demons) (The Possessed)*, 150, 189, 267n94; "The Dream of a Ridiculous Man," 151; Elder Zosima

(*Brothers Karamozov*), 73, 150–52, 155, 158, 189, 190, 191, 194–96, 199–201, 239; Ivan Karamazov (*Brothers Karamozov*), 194, 197, 198, 199, 201, 274n34; "The Land and Children," 151; Makar Dolgoruky (*The Adolescent*), 193, 199; *Notes from the Underground*, 200; Prince Myshkin (*The Idiot*), 199; the Ridiculous Man ("The Dream of a Ridiculous Man"), 196; Rodion Raskolnikov (*Crime and Punishment*), 33, 189, 194, 197, 199, 201; Sonya Marmeladov (*Crime and Punishment*), 199, 201; Ivan Shatov (*Devils*), 151; Stavrogin, Nikolai (*Devils*), 151, 194, 197; Underground Man (*Notes from the Underground*), 194, 197–99; *A Writers Diary*, 151
Douglas Spruce, 139
Dubos, René, 111; *The Wooing of the Earth*, 111
Duby, Georges, 122, 123, 124
Ducio, 135
Dufrenne, Mikel, 132, 141
Duma (Russian Parliament), 98
Duns Scotus, John, 12, 13, 15, 212
Dustin, Christopher A., 3

East, 73, 85, 125, 126, 127, 131, 135, 136, 147, 150, 154, 172, 179, 234, 269n11
Eastern Mediterranean, 172
Eastern Orthodoxy, 51, 146, 155, 196, 273n13. *See also* Orthodox Christianity
Eastern Theology, 265n49
Earth, 8, 9, 134
Eckhart, Meister, 19, 57, 61, 155, 173
Eco, Umberto, 124
Eden, 68, 89, 107, 142, 245
Egypt, 10, 34, 86, 95, 208
Egyptian Desert, 166
Egyptians, 80
Egyptian Thebaid, 72
Ehrlich, Paul R., xii
Eleusis, 3
Elijah, Prophet, 32
Eliot, T. S., 52
Ellul, Jacques, 119
Emerson, Ralph Waldo, xii, 5, 14, 40, 63, 66, 115, 131, 173, 177, 180, 241
Emory River, 68
England, 34, 39, 110, 218
English (language), 52, 169, 184, 216, 229, 235
English Romanticism, 173
Enlightenment, the, 63, 94, 97, 215
Ephraim, Elder 153, 238
Epiphany, 196
Eriugena, John Scotus, 11, 103, 173, 179, 211
Eternal Logos, 9, 12, 106, 110, 171, 209, 239, 241, 251n31
Eternal Return, 15
Eucharist, 142, 234, 235
Europe, 75, 76, 121, 183, 205
European Union, 87
European West, 47, 186
Evagrios of Pontos, 88, 167, 168, 170, 171, 180, 198, 199, 241, 242
Evdokimov, Paul, 132, 133, 136

Fall, the, 58, 82, 105, 142, 145–48, 170, 223, 244, 267n90
Feuerbach, Ludwig, 206, 224, 225, 226, 229; *Essence of Christianity*, 225
Fez, 179
First Rome, 80
Florensky, Pavel, 11, 31, 37–39, 94, 95, 98, 162, 163, 168, 183–85, 209, 235–37, 239, 243; *The Pillar and*

Ground of the Truth, 37, 116, 153, 162, 184, 235
Florida, 68, 137
Foltz, Bruce V.: *Inhabiting the Earth*, 254n20
Foucault, Michel, 17
Four Corners, x
France, 95
Francis of Assisi, St., 63, 74, 86, 155, 163, 172, 173, 188
Franciscan Order, 188
French (language), 52, 183
Friedrich, Caspar David, xii

Gadamer, Hans-Georg, 45, 116, 158
Gaia, 7
Galilei, Galileo, 11, 181, 216, 237
Gauchet, Marcel, 43, 160; *The Disenchantment of the World*, 43
George, Stefan, 18
Gerasimos, St., 188
German (language), 4, 20, 56, 192, 195, 229, 271n8
Germany, 37, 100
German Idealism, 13, 15, 29, 44, 95, 173
Gillespie, Michael, 213, 217
Giotto, 123, 135
Giving, 50
Giza, 34
Glen Canyon, 133
Gnostic(s), 128
God, xii, 5, 9–14, 16, 17, 20, 29, 34, 37, 41, 45, 49, 56, 58–60, 63–70, 73, 74, 77, 78, 82, 84, 85, 93, 94, 96, 106, 108, 113, 117–21, 123, 124, 127–32, 134, 139, 146, 147, 148, 150–54, 158, 160, 162–70, 172, 175, 176, 179, 180–83, 185–87, 189, 191, 194–97, 199, 200–3, 206, 208, 209, 211–19, 221–24, 227–30, 232, 234, 236–40, 242, 243, 245, 249n12, 250n31, 255n5, 256–57n25, 261, 265n49, 267n90, 269n11, 271n7, 274n34, 276n8, 276n20, 277n26
God's Love, 185
Goddess Nature, 218
Godhead, 38, 183
Goethe, Johann Wolfgang von, 63, 173
Golden Horn, ix, 76
Golgotha, 142
Grace, 136
Grand Canyon, 149, 261n0
Grand Inquisitor, the, 200, 274n34
Great Schism, 122, 163
Greece, 62, 87, 177
Greek (language), 6, 99, 211
Greek Atomists, 166
Greek East, 11, 74, 103, 120, 121, 122, 166, 169, 170, 174, 179, 188, 189
Greek Philosophy, 10
Greeks, 78, 96, 108, 161, 271n7
Green Corn Dance, 84
Green Patriarch, ix
Gregory of Nazianzus, St., 208, 237
Gregory of Nyssa, St., 103, 166, 167, 171, 173, 180, 208, 232, 238, 243
Gregory Palamas, St., 126, 127, 128, 129, 130, 131, 193, 210
Guattari, Felix, 227

Haar, Michel, 118
Hades, 8, 9, 142
Hagia Sophia (Church of the "Holy Wisdom of God"), 34, 79, 81, 83, 96, 105
Halchidiki Peninsula, 83
Hale, Robert, 5, 6
Half Dome, 19
Halki Seminary, 265n49
Hargrove, Eugene, 133
Harnack, Adolf, 61, 206
Hart, David Bentley, 228

Heaven, 197
Hebel, Johann Peter, 18
Hebrew (language), 47, 148
Hegel, G. W. F., 15, 19, 24, 32, 35, 43, 44, 61, 75, 95, 116, 119, 161, 228, 250n28
Heidegger, Martin, x–xiv, xvi, xvii, 1, 3, 4, 18–20, 26, 31, 35–39, 44, 50, 55–63, 67, 70, 73, 75, 79, 81, 87, 90, 92, 97, 99–101, 106, 115, 117–21, 133, 141, 145, 154, 158–60, 176–79, 183, 185, 186, 203–8, 210, 214, 216, 219, 228, 230, 234, 247n10, 255nn5–6, 256n16, 262n3, 264n42, 271n8, 276n8; *Being and Time*, 56, 59, 61, 99, 184; "Building, Dwelling, Thinking," 100; *Elucidations of Hölderlin's Poetry*, 248n10; *Der Feldweg*, 19, 133; "Introduction to The Phenomenology of Religion," 61, 255n5; "Letter on Humanism," 36, 276n8; *Mindfulness*, 256n16; "Origin of the Work of Art," 158; *The Phenomenology of Religious Life*, 255n5; "The Philosophical Foundations of Medieval Mysticism," 57; "Principle of Identity," 248n14; "The Question of Technology," 51; *Sojourns*, 1; *What Is Called Thinking?*, 18; "Why We Remain in the Provinces," 19
Hellenistic Jews, 216
Heraclitus, 7, 9, 147, 163
Heschel, Abraham, 264n38
Hesiod, 5, 131, 237; *Works and Days*, 5, 161, 184, 221
Hesychasm, 19, 63
Hetch Hetchy Valley, 133
High Sierras, xiii, 31, 173
Highland, Chris, 69
Hinduism, 82, 91
Hitler, Adolf, 95
Hiwassee River, 68
Ho Chi Minh, 95
Hobbes, Thomas, 64, 68, 256n25; *Leviathan*, 64
Hölderlin, Friedrich, xii, 18, 35, 36, 40, 50, 51, 63, 85, 106, 115, 116, 118, 121, 133, 144, 145, 155, 173, 176, 178, 231; "In Lovely Blueness . . .," 51
Holy City (earth as), 110
Holy Saturday, 10, 142
Holy Spirit, 106, 237
Homer, 14, 220; *Iliad*, 161; *Odyssey*, 161
Homeric Greeks, 7
Hopkins, Gerard Manley, 12, 40, 41, 106, 173, 200, 222; "God's Grandeur," 40, 222
Humanity, 101
Hume, David, 48, 220
Husserl, Edmund, xvii, 2, 4, 15, 26, 56, 92, 99, 224, 255n6; *Crisis of the European Sciences*, 12; *Ideas*, 92

Ibn 'Arabi, Muhyi al-Din, xiv, 182, 183, 184, 185
Ibn Rushd (Averroes), 182
Incarnation, 12, 34, 59, 73, 126, 128, 132, 142, 171, 194, 250n28; Incarnate Word, 167
International Style, 81
Invisible (the), 9, 156, 222
Isaac of Syria, St., 113, 152,163, 187, 188, 190–95, 199, 200, 249n12; *Ascetical Homilies*, 152, 190, 194
Isidore, Elder, 38
Islam, 154, 177, 182, 205, 206
Istanbul, ix, 76
Isthmus of Corinth, 76
Italy, 34, 37

Jains, 46
James, William, 4, 118, 178, 185
Janicaud, Dominique, 16
Jerusalem, 39, 61, 62, 87, 110, 164, 165, 210, 222
Jesus, 59, 195, 271n7
Jesus Prayer, xv, xvi
Job, Prophet, 221
John the Baptist, St., xiii
John of Damascus, St., 127, 128, 132
John the Evangelist, St., 245
John of Kronstadt, St., 125, 138
Jonah, Prophet, 142
Joseph the Hesychast, Elder, 238
Joseph of Volokolamsk, 125
Joyce, James: *Finnegan's Wake*, 145
Judaism, 132, 154, 177, 185, 205, 206, 225, 228, 271n7
Judean Desert, xiii
Jung, Carl, 274n28; *Modern Man in Search of a Soul*, 274n28
Jünger, G. F., 119
Justinian the Great, 77, 84

Kandinsky, Vasiliy, 95
Kansas, x
Kant, Immanuel, 15, 29, 90, 100, 101, 102, 109, 116, 138
Kartsonis, Anna, 138
Kearney, Richard, 16
Kepler, Johannes, 218, 220
KGB, 183
Kidron Valley, 87
Kierkegaard, Søren, 15, 61, 116, 117
Kiev, 78, 105
Kingdom, the, 136
Kingdom of Sisyphus, 43
Kisiel, Theodore, 119
Klein, Jacob, 7, 249n23; *Lectures and Essays*, 249n23
Kohák, Erazim, 45, 92, 93, 94, 107; *The Embers and the Stars*, 92
Kojeve, Alexandre, 95
Kontakion, 128
Kontoglou, Photios, 193
Koyré, Alexandre, 95, 217, 218; *From Closed World to Infinite Universe*, 217
Krishna, 122
Kristeva, Julia: *In the Beginning Was Love: Psychoanalysis and Faith*, 150
Kropotkin, Peter, 95
Krutch, Joseph Wood, 115
Kuhn, Thomas, 220

"La Camaronera," xv
Lamb of God, 39
Late Antiquity, 80, 82, 85
Latin Middle Ages, 130, 154, 188, 204, 205, 206
Latin West, xiii, 11, 12, 15, 49, 86, 103, 121, 122, 124, 128, 170, 179, 187, 188, 193, 206, 231, 235
Lao Tzu, 108, 131
Law, 170
Lawrence, D. H., 173
Leibniz, Gottfried Wilhelm, 184
Lenin, Vladimir, 95, 98
Lent, 128
Leonardo da Vinci, 237
Leopold, Aldo, 115, 151, 173; "The Land Ethic," 187
Levinas, Emmanuel, 16, 17, 224, 228, 229, 277n26; *Difficult Freedom: Essays on Judaism*, 277n26
Levy-Bruhl, Lucien, 45
Lewis, C. S., 240
Libri Carolini, 12, 83
Light, 156
Liturgy, 84, 234

Liturgy of Holy Saturday, 197
Liturgy of St. James, 196
Locke, John, 65, 213, 216, 218, 220
London, 95
Logos, 12, 82, 83, 88, 126, 128, 129, 141, 166, 168, 169, 171, 181, 194, 216, 220
Lopez, Barry, 14, 40, 50, 63, 94, 104, 115, 131, 173, 177; *About This Life: Journeys on the Threshold of Memory*, 131
Lord, 68, 106, 125, 127, 182, 194, 197, 199, 221, 222, 235, 237, 245
Lord's Prayer, 201
Lossky, Vladimir, 130
Louvre, 233
Love, 232
Lucretius, 2, 254n13, 275n5; *De Rerum Natura*, 275n5
Luther, Martin, 20, 55, 57, 58, 59, 60, 61, 63; *Heidelberg Dissertation*, 20, 57

Macedonia, 83
Mahabharata, 122
Maimon, Moshe ben (Maimonides), 210
Malevich, Kazimir, 95
Manichaeans, 128
Manoussakis, John Panteleimon, 16, 250n31; *God After Metaphysics*, 250n31
Mao Zedong, 95
Marion, Jean-Luc, 44, 50, 93, 179, 213, 214, 224, 226, 228
Marks, Steven, 95; *How Russia Shaped the Modern World*, 95
Martin of Tours, St., 85, 269n11
Marx, Karl, 6, 44, 75, 90, 98, 99, 100, 101, 192
Marxism, 93
Mary, Virgin, 9, 140, 142, 150
Masaccio, 135
Mass, 170
Mathew, Gervase, 240; *Byzantine Aesthetics*, 240
Matins, 128
Maximos the Confessor, St., 12, 58, 83, 103, 128–31, 147, 158, 169, 170–74, 179–84, 194, 199, 210, 211, 240–42; *Mystagogia*, 130
Maya (Illusion), 82, 108
Mecca, 179
Megara, 76
McKibben, Bill, 39, 52; *The End of Nature*, 39, 52
McPhee, John, 71
Medieval Scholasticism, 13
Mediterranean, 46
Merleau-Ponty, Maurice, 45, 99; "Nature," 215
Merton, Thomas, 174
Mesoamericans, 80
Messian, Olivier, xii
Miami, xv
Middle Ages, 120, 121, 122, 123, 156, 205, 206, 211
Middle East, 131
Middle Platonism, 165, 240
Midgley, Mary, 253n5; *Beast and Man*, 253n5
Milbank, John, 16, 17, 27
Milosz, Czeslaw, 47, 254n15; "Advice," 47
Milosz, O. V. de L., 45, 254n15; *The Arcana*, 45
Minoans, 80
Mistra, 125
Momaday, Scott, 131, 145; *The Way to Rainy Mountain*, 145
Moses, 11, 31, 126, 127, 134, 147, 167, 169, 222, 242, 277n26

Mothersill, Mary, 27
Mount Athos, 3, 74, 83, 86, 126, 172, 188, 237, 238
Mount Ktaaden, 134
Mount of Olives, 87
Mount Sinai, 35, 126, 134, 277n26
Mount Tabor, 34, 65, 126
Muir, John, xii, xii, xiv, 14, 29, 31, 40, 49, 58, 63, 64, 66–74, 94, 104, 115, 131, 139, 143, 155, 156, 173, 174, 190, 200, 217, 220, 241, 247n7, 257n27, 261, 267n90; *The Cruise of the Corwin*, 69; "The Grand Canyon of the Colorado," 261; *Thousand Mile Walk to the Gulf*, 68
Mussolini, Benito, 95

Nabhan, Gary Paul, 14
Naddaf, Gerard: *The Greek Concept of Nature*, 275n5
Naess, Arne, 25, 26, 91, 105; "Deepness of Questions and the Deep Ecology Movement," 25
Naipaul, V. S., 104
Nash, Roderick, 257n27
Nasr, Seyyed Hossein, 22, 95, 155, 156
Nativity of Christ, 9, 10
Nature, 47, 84, 93, 176, 215, 219
Navahos, x
Nazi Germany, 37
Neo-Platonism, 66, 95; Neoplatonists, 208
Neo-scholasticism, 57
Neuhäuser, Rudolf, 190
New Age, 95
New Jerusalem, 79, 110, 202
New Prometheus, 208
New Testament, 67, 142, 216, 221
"New Testament Christian," 60
Newton, Isaac, 110, 181, 213, 216, 218, 219, 220
Nicholson, Marjorie Hope, 103
Nietzsche, Friedrich, xii, xiii, 15, 21, 23, 31–35, 67, 71, 74, 75, 81, 85, 94, 116, 117, 150, 153, 154, 159–62, 164, 166, 172, 180, 184, 192, 193, 206–9, 224, 226, 229, 233–36, 241, 242; *The Birth of Tragedy*, 32; *Beyond Good and Evil*, 159; *The Gay Science*, 33, 159; *The Twilight of the Idols*, 226; *Thus Spoke Zarathustra*, 23, 33
Nijinsky, Vaslav, 95
Nobel Laureate, 254n15
Nobodaddy, 219, 227
Norman Conquest, 12
North America, xv
Northern Africa, 131
Northern Syrian Highlands, 86
Novalis, 88
Novgorod, 105

Oates, Joyce Carol, 40
Ockham, William of, 13,15,121, 122, 124, 126, 173, 188, 212, 213
Oelschlager, Max, xi, xiii, 115, 262n4; *Caring for Creation*, x
O'Keeffe, Georgia, xii
Old Byzantium, 122
Old Night, 134
Old Testament, 34, 59, 66, 127, 142
Onasch, Konrad, 138
One, the, 46, 79
Optina Monastery (Russia), 200
Origen of Alexandria, 166, 167, 170, 171, 180
Orthodox Christianity, 127, 190, 191, 234, 241, 265n52. *See also* Eastern Orthodoxy

Orthodox Church, 10, 122, 239. *See also* Church: Eastern Orthodox
Orthodox East, xv, 74, 86, 98, 172, 237
Ostia, 211
Other, the, 16
Ott, Hugo, 119
Otto, Walter, 6, 7
Ottoman Empire, 122
Ouspensky, Leonid, 146
Oxford, 126

Paisios, Elder, 74, 172, 188, 237; *Epistles*, 235
Palestine, 10, 11, 86, 208
Panagia, 238
Pantheon, 81
Papadiamandis, Alexandros, 87
Paradise, 47, 86, 142, 145, 146, 148, 149, 151
Paris, 98, 254n15
Parmenides, 4, 7, 8, 9, 14, 18, 166; *On Truth*, 8
Pascal, Blaise, 61
Pascha (Easter), 238
Paul, St., 57, 59, 60, 61, 168, 231, 233, 245, 255n5
Paul the Silentary, 84
Peace, 65
Peloponnese, 76
Pennington, Basil, 174
Pentateuch, 242
Perl, Eric D., 270n22; *Theophany*, 250n27, 270n22
Persians, 108
Peter of Damascus, St., 239
Petri, Elfriede, 255n6
Phanar "Lighthouse", ix
Phidias, 223, 227, 230
Philo of Alexandria, 128, 165, 206
Philokalia, 4, 38, 199, 236, 249n12
Pickstock, Catherine, 16, 17
Pinchot, Gifford, 68
Pindar, 5
Piraeus, 8, 87
Plato, 4, 8, 9, 10, 14, 18, 28, 29, 30, 34, 46, 63, 96, 103, 106, 164–66, 178, 182, 201, 206, 211, 224; *Laws*, 220; *Phaedrus*, 8, 10, 164, 184; *Republic*, 8, 10; *Symposium*, 4; *Timaeus*, 34, 164
Platonism, 15, 63, 67, 71, 96, 106, 131, 153, 159, 164, 206, 207
Platonist(s), 167
Plotinus, 4, 46, 77, 103, 166
Plutarch, 161
Pöggeler, Otto, 56
Popescu, Dimitru, 265n49
Porphyrios, Elder, 237, 238; *Wounded by Love*, 116, 232
Porter, Eliot, 104
Portofino, 31
Poseidon, 87
Pound, Ezra: *Cantos*, 145
Powell, John Wesley, 104, 149
Prayer, Christian, 159
Procopius, 84
Problem, 82
Pre-Socratics, 14, 220
Protestant, 55, 116
Protestantism, 102, 255n6
Psalmist, 58, 223, 242
Psalter(s), 46, 124
Pueblo Indians, 274n28

Radical Orthodoxy, 16
Rapallo, 31
Ravenna, 122
Reformation, 63
Reformer, 61
Renaissance, 13, 81, 123, 125, 195, 241; Renaissance Humanism, 217

Renaissance West, 46, 138
Representatives of Twenty Monasteries, 238
Resurrection, 10, 238
Rilke, Rainer Maria, 18, 109
River Isis, 237
River Jordan, 142
Road to Emmaus, 168
Rock, Stella; *Popular Religion in Russia: "Double Belief" and the Making of an Academic Belief*, 273n13
Roman Catholicism, 273n13
Roman Empire, 72, 78
Roman Philosophy, 275n5
Romans, 59, 161, 233
Romanticism, 131
Romantics, 209
Rome, 1, 76, 80, 81, 122, 155
Rublev, St Andrei, 137, 264n34
Ruskin, John, 140, 141
Russia, xiii, 34, 37, 73, 74, 78, 94, 183, 184, 274n34; 1905 Revolution, 98; Czarist Russia, 100; Post-Soviet Russia, 100; Russian Revolution, 98; Soviet Era, 94; Stalinist Russia, 37
Russian East, 174
Russian Orthodoxy, 105, 273n13
Russian Pilgrim, xvi, 239

Sacred, the, 22, 228
San Francisco, 89
Sandbank, S., 66
Santa Fe Railroad, x
Savannah, 68
Savior, the, 240
Satan, 218
Sax, Joseph, 69
Scheler, Max, 45, 46, 48
Schelling, F. W. J., 101, 103, 173
Schiller, Friedrich, 6
Schleiermacher, Friedrich, 16, 255n6
Schnieper, Annemarie, 138
Scholasticism, 11, 15, 63, 207, 208; Late Scholasticism, 185
Scholastics, 71, 204, 209, 230
Schopenhauer, Arthur, 138
Scotland, 71
Scottish Lowlands, 67
Scotus, John Duns, 56, 188
Scully, Vincent, 80, 81, 83, 84, 86; *Architecture: The Natural and the Manmade*, 80
Scylla and Charybdis, 231
Sea of Marmara, 76
Second Nicene Council, 83
Sepänmaa, Yrjö, 262n4
Seraphim of Sarov, St., 172, 188
Sermon on the Mount, 194, 195, 221
Shakespeare, William, 95
Sheperd Paul, 111
Sherrard, Philip, 155; *The Sacred in Life and Art*, 143
Shestov, Lev, 95, 98
Sierra Club, xi, 104
Siberia, xvi
Silko, Leslie Marmon, 145
Sinai, 11, 222
Slavonic, 191
Smith, James K. A., 16
Snyder, Gary, 40, 49, 63, 115, 154, 156, 173
Solovyov, Vladmir, 31, 33, 34, 35, 37, 95, 103
Son, the, 140
Song of the Earth, 111
Song of the Seraphim, 221
Sophia, 34, 38, 103, 105, 106, 107, 109, 110, 111, 183, 201, 216, 220, 261n47; Holy Wisdom of God, 46
Sophronios, St., 196

Soviets, 273n13
Spinoza, Baruch, 91, 216, 217, 218
Spinozism, 90
Spirit, 29, 142, 165
Socrates, 184
Stanislavski, Constantin, 95
Steiner, George, 93, 95; *Nostalgia for the Absolute*, 93
Stithatos, Nicholas, 38, 199, 236
Stoicism, 90
Stoics, 103, 147, 163
Stravinsky, Igor, 95
Sufism, 183, 197
Sun, 8
Switzerland, 33, 98
Symeon the New Theologian, St., 146, 210
Symeon the Stylite, St., 86, 146
Symeon Stylites the Younger, St., 86
Syria, 11, 86
Syrian Deserts, 166

Tantalus, 23
Tao, 108
Taoism, 185, 234
Taos Pueblo, 84, 274n28
Tarkovsky, Andrei, 263n34; *Andrei Rublev*, 263n34
Tauler, Johannes, 173
Taylor, Charles, 1, 17, 250n28; *A Secular Age*, 1, 250n28
Teale, Edwin, 115
"Temple of the Moon," 80
Temple Mount, 87
Tennessee, 68
Teotihuacán, 80
Teresa of Ávila, St., 61, 185
Tertullian, 61
Tinker Creek, 144, 173
Titan, 134
Thales, 233, 235
Themistius, 76
Theodore, St., 242, 244
Theodulf of Orléans, 13
Theotokos, 136
Thessaloniki, 3, 87; St. Demetrius Cathedral, 87
Thomas, Lewis, 144
Thoreau, Henry David, xii, 14, 31, 40, 49, 63, 66, 93, 104, 106, 115, 131, 134, 144, 149, 155, 173, 177, 180, 190, 219, 241; *Walden*, 144; *A Week on the Concord and Merrimac Rivers*, 144, 145
Thou, 141
Tikhon of Zadonsk, St., 150
Tolkien, J. R. R., 240
Tolstoy, Leo, 28, 29, 95, 236
Traherne, Thomas, xiv, 20, 58, 63, 64, 65, 66, 67, 68, 69, 70, 71, 72, 73, 256–57n25; *Centuries Meditation*, 64; *Christian Ethicks*, 64
Trakl, Georg, 18
Trans-Caucasus, 37
Transfiguration, 34, 126
Trinity, 220; Trinitarian Life, 261n47
"Triumph of Orthodoxy," 128
Troy, 220
Tuan, Yi-Fu, 143
Tuolomne Valley, 69
Turkey, 95, 265n49
Turner, J. M. W., xii
Twain, Mark, 144
Tyutchev, Fyodor, 88

Urizen, 213, 219
Utah, 208

Van Buren, John, 56, 119
Van Gogh, Vincent, 230

Vanheeswijck, Guido, 17
Vasileios, Archimandrite, 136
Vattimo, Gianni, 16, 61
Vedanta, 236
Velimirovich, St. Nikolai, 238
Vico, Giambattista, 75
Virgil, 131, 209; *Georgics*, 10, 184, 221
Visible, the, 156
Vladimir, Grand Prince of Kiev, 78, 96
Von Baader, Franz, 103

Wagner, Richard: *Twilight of the Gods*, 226
Walden Pond, 173
Ward, Graham, 16, 17
Watson, Robert, 256n25; *Back to Nature: The Green and the Real in the Late Renaissance*, 257n25
Way of Appearing, 14
Way of Truth, 14
Weber, Max, 6, 85, 91, 161, 188, 208; *Economy and Society*, 91
Wedding, 139
West, the, 34, 73, 74, 81, 85, 87, 94, 95, 97, 98, 107, 118–20, 122, 125, 127, 128, 132, 135, 145, 154–56, 163, 172–74, 179, 182, 185, 186, 188, 189, 192, 209, 210, 213, 229, 234, 241, 268n105
Western Africa, xv
Western Europe, xiii, xv, 11, 38, 42, 63, 74, 80, 120, 122, 125, 151, 155, 190, 197, 198, 205
Western Middle Ages, 15, 119
Western Renaissance, 250n28
Weston, Edward, 104
Whitehead, Alfred North, 91, 115
Whitman, Walt, 144
Will, 29
Williams, Dennis, 267n90; "John Muir, Christian Mysticism, and the Spiritual Value of Nature," 267n90
Wisdom, 7, 211, 261n47
Wittgenstein, Ludwig, 20
White, Lynn Jr. xiv, 46, 74, 75, 119, 120, 121, 123, 124, 154, 160, 162, 187–89, 200, 202, 204–7, 213, 233, 235; "The Historical Roots of our Ecologic Crisis," 74, 119, 187; *Medieval Technology and Social Change*, 120
Wordsworth, William, xii, 155, 173
Worringer, Wilhelm, 81
Worster, Donald, 257n27; *A Passion for Nature: The Life of John Muir*, 257n27

Yannaras, Christos, 2, 4, 224, 229, 230, 231; *Person and Eros*, 229; *Variations on the Song of Songs*, 215
Yeats, William Butler, 77, 78, 79; "Sailing to Byzantium," 77, 119; *A Vision*, 77; *The Works of William Blake*, 77
Yosemite Valley, xiii, 69, 133

Zarathustra, 32, 33, 35, 74, 150, 159, 192, 207, 226
Zeigler, Joanna E., 3
Zeus, 7, 8

groundworks |

ECOLOGICAL ISSUES IN PHILOSOPHY AND THEOLOGY

Forrest Clingerman and Brian Treanor, *Series Editors*

Interpreting Nature: The Emerging Field of Environmental Hermeneutics
Forrest Clingerman, Brian Treanor, Martin Drenthen, and David Utsler, eds.

The Noetics of Nature: Environmental Philosophy and the Holy Beauty of the Visible
Bruce V. Foltz

Environmental Aesthetics: Crossing Divides and Breaking Ground
Martin Drenthen and Jozef Keulartz, eds.

The Logos of the Living World: Merleau-Ponty, Animals, and Language
Louise Westling